The Entropic Basis of Cosmos Theory

Spontaneous Evolution of the Cosmos
HyperSphere Universe Coordinates

Stephen Blaha Ph. D.
Blaha Research

MMXXV

ISBN: 979-8-9986071-9-6

Rev. 00/00/01 December 20, 2025

To Margaret

Some Other Books by Stephen Blaha

SuperCivilizations: Civilizations as Superorganisms (McMann-Fisher Publishing, Auburn, NH, 2010)

Unification of God Theory and Unified SuperStandard Model THIRD EDITION (Pingree Hill Publishing, Auburn, NH, 2018).

The Exact QED Calculation of the Fine Structure Constant Implies ALL 4D Universes have the Same Physics/Life Prospects (Pingree Hill Publishing, Auburn, NH, 2019).

Passing Through Nature to Eternity ProtoCosmos, HyperCosmos, Unified SuperStandard Theory (Pingree Hill Publishing, Auburn, NH, 2022).

HyperCosmos Fractionation and Fundamental Reference Frame Based Unification: Particle Inner Space Basis of Parton and Dual Resonance Models (Pingree Hill Publishing, Auburn, NH, 2022).

The Cosmic Panorama: ProtoCosmos, HyperCosmos, Unified SuperStandard Theory (UST) Derivation (Pingree Hill Publishing, Auburn, NH, 2022).

God and and Cosmos Theory (Pingree Hill Publishing, Auburn, NH, 2023).

Newton's Apple is Now The Fermion (Pingree Hill Publishing, Auburn, NH, 2023).

Cosmos Theory: The Sub-Particle Gambol Model (Pingree Hill Publishing, Auburn, NH, 2023).

Cosmos-Universe-Particle-Gambol Theory (Pingree Hill Publishing, Auburn, NH, 2024).

Structure and Dynamics of Cosmos Theory and the Unified SuperStandard Theory (Pingree Hill Publishing, Auburn, NH, 2024).

Black Holes, White Holes, and Superluminal Starship (Pingree Hill Publishing, Auburn, NH, 2025).

The Sequence Theory Foundation of Cosmos Theory and Gambol Theory (Pingree Hill Publishing, Auburn, NH, 2025).

BEYOND TIME AND SPACE Fermion Strata & Consistency Conditions for Sequences, Cosmos Spaces and Gambols (Pingree Hill Publishing, Auburn, NH, 2025).

The C o s m o s as a Physics Organism (Blaha Research, Auburn, NH, 2025)

The Root of Reality (Blaha Research, Auburn, NH, 2025)

Fractal Cosmos (Blaha Research, Auburn, NH, 2025)

Fractal Cosmos II (Blaha Research, Auburn, NH, 2025)

Fractal Cosmos III (Blaha Research, Auburn, NH, 2025)

Fractal Cosmos IV (Blaha Research, Auburn, NH, 2025)

Available on bn.com, Amazon.com, Amazon.co.uk and other international web sites as well as at better bookstores.

CONTENTS

FIGURES and TABLES

Introduction

Cosmos Theory gives a detailed approach to understanding the origin of the Cosmos. It suggests it all happens at one instant where the instant is "forever" – not in the infinite past – not in the infinite future – not in time. Eternally instantaneous. How could such a thing be envisioned? The book presents the process broken down into a series of Physical stages from an initial Mathematical point to the full Cosmos. The book also develops the origin of the Cosmos from a single dimension that grows to be an almost innumerable set of dimensions.

In both scenarios there is an instantaneity. But the book must describe it consecutively. The great question is: What Power forces the sequences of events depicted here? Physically it becomes clear that there must be a governing Physics Principle – Entropy – that defines the way a reaction (interaction) must proceed. This principle is seen in Chemistry – reactions have directions generally – from a small entropy to a larger entropy based on the complexity (chaos factor) of the initial and final stages of the Chemistry reaction. Here the reaction is the growth process from the infinitesimal to the large.

This book considers the growth of the Cosmos (instantaneous as it is) entropically and sees that it proceeds entropically: the growth of Cosmos spaces, the growth of universes, and the growth of parts thereof.

The book leads to the great question: Why did the growth initiate? One might also ask: Why the starter: one point or one dimension? It leads to a possible answer: the existence of an Agent who has two roles. One role is to be the unchanging initiator – the Prime Mover – that "starts" it (although there is no start time.) The Agent is the initiator of "what is" where the "is" (the Cosmos structure) is unchanging. Thus we see the initiator must also be unchanging in this role.

As the "creator" of child universes, which must proceed according to the Laws of Physics, we see a dynamic role for the Agent within the Parent universe and within child universes.

Why does the Agent act dynamically? The only apparent answer is to have the scope and involvement of dynamic activity within universes.

This provocative book ends gracefully at this point.

1. Cosmos Entropy Rules

The goal of this chapter and the following chapters is to show that the processes that generate the Cosmos structure of spaces have an entropy that increases from stage to stage. As a result the processes are allowed by the Second Law of Thermodynamics. Once started, they can proceed to completion without requiring the action of an external agency such as a Prime Mover.

We necessarily must describe the entropic processes consecutively. However one must realize the processes occur simultaneously. Due to the lack of a time they coexist instantaneously – not at a particular point in time. We are aided in the examination of processes from an entropic view by the use of the Boltzmann statistical formulation of entropy. We determine an entropy by counting microscopic states of sets of dimensions. We can then compare entropies at the beginning and the end of processes.

1,1 Entropy

Physical and Chemical processes follow the Second Law of Thermodynamics. Entropy cannot decrease with time by the Second Law of Thermodynamics. The entropy of most processes tends to automatically increase with time. In our recent books we have shown a detailed evolution of Cosmos Theory from a single mathematical point. While the description of this evolution is mathematically compelling the question still remains: Why do its steps take place? If we consider Darwin's Theory of evolution we see he found a reason for evolution: Survival of the Fittest.

We will show that the evolution of Cosmos Theory also has a compelling reason: the Increase of Entropy.

It is common to visualize entropy starting at a low value and then increasing to a larger value for a Physical or Chemical system. We associate high entropy with a maximal disordered system and low entropy with an ordered system. We will see the evolution of the features of Cosmos Theory seems to happen *automatically* in accord with the growth of entropy (by the Second Law). It could not be otherwise.

1.2 Entropy of a Fractal Curve

We can define the *Entropy* S_e of a fractal curve at each step n of the construction process as a product of a constant (that we will take to be one) and a measure of the number n of possible microstates (arrangements) of a system.[1] As the number of possible arrangements increases the entropy increases. We view the initial state as

[1] Our definition of entropy is based on Boltzmann's definition of entropy. The author feels a certain kinship with Boltzmann who was also the recipient of a colleague's unwarranted, extreme criticism and harassment for his – now accepted – Atomic Theory of Statistical Mechanics – in the 19th Century when the atomic theory of matter was still unaccepted by most Physicists .

having low entropy because of the low number of possible arrangements (microstates). As the number of mictrostates increases the entropy increases to a maximum.

The entropy S_e is defined to be

$$S_e = k_C \ln_2(n) \tag{1.1}$$

where k_C is a dimensionless constant corresponding to the Boltzmann constant k_B. Note S_e is also dimensionless. We will tentatively set $k_{Cosmos} = 1$. We choose the logarithm of base 2 with a view towards using it for Cosmos spaces.

Turning now to the Cosmos Fractal Curve (the Hilbert fractal curve) we start with the n = 1 step in the construction (after the generation of the unit line segment). The entropy is

$$S_e(1) = k_C \ln_2(1) = 0 \tag{1.2}$$

The nth line construction step incorporates the sum of line segments t_n

$$t_n = \sum_{k=1}^{n} 2^k = 2^n - 1 \tag{1.3}$$

We form a measure of the number of microstates (arrangements) of the construction step linear segments of length 2^n realizing they may be reordered arbitrarily:

$$m_n = 4/(1 + t_n^{-1}) = 4\, t_n/(1 + t_n) = 4(1 - 2^{-n}) \tag{1.4}$$

resulting in

$$m_{n=\infty} = 4$$
$$m_{n=1} = 2$$

and an entropy of

$$S_e(m_n) = k_C \ln_2(m_n) = \ln_2(m_n) \tag{1.5}$$

$$S_e(m_{n=\infty}) = 2$$
$$S_e(m_{n=1}) = 1$$

just as the fractal curve begins with a dimension 1 line segment and grows to a dimension 2 fractal curve. Within the construction process the entropy increases as expected.

Eq. 1.3 expressed in the form

$$t_n^{-1} \sim 1/(e^{h\omega/kT} - 1) \tag{1.6}$$

can be interpreted in a statistical manner as the Planck Radiation Law (Distribution), which Einstein showed could be based on statistical considerations of transitions between radiation states. In this form

$$2^n \sim e^{h\omega/kT} \tag{1.7}$$

implying that a dimension array of size 2^n has a temperature

$$kT = h\omega/\ln(2^n) \tag{1.8}$$

The Cosmos Fractal Curve construction steps have been matched with Cosmos spaces in our previous work. Cosmos spaces have an associated energy that we have seen in our Consistency Condition studies:

$$E = h\omega \sim 2^n \tag{1.9}$$

Together with eq. 1.8 we find

$$kT \sim 2^n/\ln(2^n) = (2^n/n)/\ln(2) \tag{1.10}$$

The entropy and temperature grow with the increase in construction steps.

Following a similar approach as eq. 1.5 for fractal curves we find the entropy of the Planck distribution is

$$\begin{aligned} S_e(a_n t_n^{-1}) &= k_C \ln_2(a_n t_n^{-1}) = \ln_2(a_n/(e^{h\omega/kT} - 1)) \\ &= -k_C \ln_2(e^{h\omega/kT} - 1) + k_C \ln_2(a_n) \end{aligned} \tag{1.11}$$

using eq. 1.6 where a_n is the factor in the Planck distribution. Note the entropy is always defined up to a constant. Consequently we may take the entropy of the Planck distribution to be

$$S_e(a_n t_n^{-1}) = k_C \ln_2(1/(e^{h\omega/kT} - 1)) \tag{1.12}$$

The Planck distribution entropy increases as the energy ω decreases. As the energy decreases the number of (photon) microstates increases (infrared catastrophe) causing the increase in entropy.

1.3 Entropy of a Dimension Array

We now consider the entropy of dimension arrays of the Cosmos spaces spectrum of 10 HyperCosmos and 10 2nd Kind HyperCosmos spaces. The largest *Physical* space has 18 dimensions. The least *Physical* space has 0 dimensions.[2]

Each Cosmos space has a dimension array. The dimension array is its only characteristic. It embodies space-time and symmetry groups. A dimension array is a set of dimensions without numeric values. Dimensions are merely place holders. They map to a space-time, and to the dimensions of irreducible representations of the space's symmetry groups. They also map to other sets such as the set of fundamental fermions associated with the space when a universe is created of the type of the Cosmos space.

As we see in detail later we may view a dimension array as created dimension by dimension from an initial single dimension.

A dimension array construction starts with a single dimension and low entropy where $S_{eLOW} = \ln_2(1) = 0$ and proceeds to high entropy where the dimension array is

---

[2] Similar comments apply for Limos spaces where the dimension is less than zero.

completely filled and $S_{eHIGH} = \ln_2(d_{dr}) = r + 4$ where $d_{dr} = 2^{r+4}$ is the size of the dimension array. (See Figs. 1.1 – 1.3.) We choose to use the base 2 logarithm $\ln_2$ since Cosmos spaces are measured usually in powers of two.

We view the initial one dimension state as having low entropy because the number of possible arrangements (microstates) is one. The value of S_{eHIGH} is the *fractal dimension* of the Cosmos space[3] of dimension r. (For a space with strata and gambol fermions $S_{eHIGH} = \ln_2(2^{r+7}) = r + 7$.)

We view the construction of the dimension array as the result of applying transformations to the dimension array that increase the number of dimensions one by one. At each step the dimension array size d(n) (the number of non-zero components) undergoes:

$$d(n+1) = Td(n) \tag{1.13}$$

where T is a combined General Relativity – Symmetry Group transformation. At each step the entropy of the dimension array increases to

$$S_e(n+1) = k_C \ln_2(n+1) \tag{1.14}$$

At the maximum value of the number of dimensions $d_{nr} = 2^n$ for the dimension array the entropy of the dimension array is

$$S_e(d_{nr}) = k_C \ln_2(2^n) = k_C n = n \tag{1.15}$$

As the number of dimensions increases, the dimension array entropy increases reaching a maximum at the full dimension array for a given space. The number of possible arrangements of the full set of dimensions is at a maximum – corresponding to maximal dimension array entropy and the space's fractal number.

1.4 Entropy Production Mechanism for a Dimension Array

In Blaha (2023a) and earlier work[4] ranging back to the 1970's we developed the mechanism for reducing – and growing – dimension arrays. We showed that a combination of General Relativistic[5] and Symmetry Group transformations can take a vector with d_{dr} components from having only one non-zero component (dimension) to a vector with d_{dr} non-zero components (dimensions). In the study of Cosmos dimensions, their numerical value is not relevant. What is relevant is whether they are zero or non-zero. Non-zero components map to irreducible symmetry group dimensions and to fundamental fermions.

---

[3] Described in Blaha (2025i).

[4] S. Blaha, Phys. Rev. **D17**, 994 (1978); S. Blaha, "The Local Definition of Asymptotic Particle States", IL Nuovo Cimento **49A**, 35 (1979).

[5] The above papers showed that general relativistic transformations including transformations between accelerating reference frames cause transformations between positive and negative energy fermion raising and lowering operators. This feature carries over into effects on dimension array components as discussed in the above papers.

The form of the transformation T of a $d_{dr} = 2^{r+4}$ component vector[6] d_0 to another vector d_1 is

$$d_1 = Td_0 \qquad (1.11)$$

where T is a d_{dr} by d_{dr} array. (After a transformation the resulting d_{dr} component vector can then be put into rectangular form. See Blaha (2023a) for more details.)

We now consider the sequence of transformations that takes a d_{dr} component vector with only one non-zero component to a d_{dr} component vector filled with non-zero values. As the d_{dr} component vector gets filled – step by step – the entropy of the corresponding dimension array, as defined above, grows. Thus we see a growth from a low entropy to a high entropy. This growth, which may be viewed as happening for all Cosmos spaces' dimension arrays, is part of the process of generating the set of Cosmos spaces from a single point (which might be viewed as a primeval dimension.) Given a space and its empty (except for one non-zero dimension) dimension array we generate a filled dimension array entropically – effectively automatically.

This growth pattern applies to all dimension arrays of all Cosmos Theory spaces. In the next chapters we consider our Cosmos Theory unification spaces and show that all dimensions of all spaces grow from one primeval dimension in the 180 dimension PenUltimate Cosmos space. This growth is instantaneous and governed by entropy. Entropic growth allows instantaneous, determined growth from one point (dimension). It is usually automatic just like chemical reactions – it proceeds automatically – entropically.

[6] This discussion also applies when we introduce fermion strata and gambol fermions. Then $d_{dr} = 2^{r+7}$.

THE HYPERCOSMOS SPACES SPECTRUM

Blaha Space Number N_s	Cayley-Dickson Number n_c	Cayley Number Cn d_c	Dimension Array column length $d_{cd} = d_{cr}$	Dimension Array Size $d_{dN}=d_{dr}$	Space-time-Dimension r	CASe Group $su(2^{r/2},2^{r/2})$ CASe
0	10	1024	2048	2048^2	18	su(512,512)
1	9	512	1024	1024^2	16	su(256,256)
2	8	256	512	512^2	14	su(128,128)
3	7	128	256	256^2	12	su(64,64)
4	6	64	128	128^2	10	su(32,32)
5	5	32	64	64^2	8	su(16,16)
6	4	16	32	32^2	6	su(8,8)
7	**3**	**8**	**16**	$\mathbf{16^2}$	**4**	**su(4,4)**
8	2	4	8	8^2	2	su(2,2)
9	1	2	4	4^2	0	su(1,1)
10	0	1	2	2^2	-2	su(0,0)
11	-1	½	1	1	-4	su(-1,-1)

Figure 1.1. The HyperCosmos space spectrum.

THE Limos SECTOR SPACES SPECTRUM

Blaha Space Number $N = O_s$	Cayley-Dickson Number n	Cayley Number d_c	Row Size d_{cr}	Dimension Array Size $d_{dN}=d_{dr}$	Space-time-Dimension r	CASe Group $su(2^{r/2},2^{r/2})$ CASe
10	0	1	2	2^2	-2	U(1)
11	-1	½	1	1^2	-4	U(½)
12	-2	¼	½	$½^2$	-6	U(¼)
13	-3	1/8	¼	$¼^2$	-8	U(1/8)
14	-4	1/16	1/8	$1/8^2$	-10	U(1/16)
				•		
				•		
				•		
∞	−∞	0	0	0	−∞	U(0)

Figure 1.2. The Cosmos Theory Limos spaces spectrum for Gambol Theory. See Blaha (2023e).

HYPERCOSMOS OF THE SECOND KIND SPACES SPECTRUM

Blaha Space Number	Cayley-Dickson Number	Cayley Number	Dimension Array size	Space-time-Dimension	CASe Group $su(2^{r/2},2^{r/2})$
$N = O_s$	n	d_c	$d_{dN2} = d_{dr2}$	r	CASe
0	10	1024	1024 × 2048	18	su(512,512)
1	9	512	512 × 1024	16	su(256,256)
2	8	256	256 × 512	14	su(128,128)
3	7	128	128 × 256	12	su(64,64)
4	6	64	64 × 128	10	su(32,32)
5	5	32	32 × 64	8	su(16,16)
6	4	16	16 × 32	6	su(8,8)
7	**3**	**8**	**8 × 16**	**4**	**su(4,4)**
8	**2**	4	4 × 8	2	su(2,2)
9	**1**	2	2 × 4	0	su(1,1)

Figure 1.3. Part of the HyperCosmos of the Second Kind space spectrum.

2. Generating All Cosmos Spaces from One Dimension

We now describe how to generate all Cosmos spaces from one dimension in the 180 dimension Penultimate Unification space.[7] First we define a Fundamental Reference Frame (FRF) as a limiting (FRF) dimension array with one component (primeval) dimension that is the result of a transformation(s) from a conventional dimension array filled with non-zero component dimensions. Then we consider the inverse transformation that takes a one non-zero dimension component FRF to a filled dimension array. We will use this entropic process in chapter 3 to define entropic transitions that are used to entropically create the spaces' structure of Cosmos Theory.

2.1 The Fundamental Reference Frame (FRF) as a "Rest Frame" for Reference Frames

A common feature of calculations in Physics is to choose a momentum reference frame which facilitates computation. Common choices are the rest frame and the center of mass frame. In this chapter we describe our Fundamental Reference Frames (FRF) (actually an infinite set of such frames) that plays a similar role for coordinate reference frames. We now describe how to determine a Fundamental Reference Frame for each of the Cosmos spaces.

There are two types of FRFs that are relevant for unification and for our entropy studies:

1. An FRF that has only one non-zero dimension. All other dimensions are zero.
2. An FRF that has a set of non-zero dimensions amounting to the length of a dimension array column: $d_{cr} = d_{dr}^{1/2}$. (See Fig. 1.1.)

Each type can be generated from a dimension array by applying a combined General Relativistic – Internal Symmetry transformation to the dimension array made into a vector as we saw in chapter 1 and in Blaha (2023a) and earlier books.

The FRF transformations were described in these books as well as our papers in the 1970's based on the close analogy of dimension arrays, Dirac γ-matrices, and fermion raising and lowering operators. The numeric value of dimensions is only relevant as to whether they are zero or non-zero. Consequently we may view transformation components as simply c-numbers. Non-zero dimension array components map to fermions and also map to irreducible symmetry group dimensions.

[7] Much of this chapter is based on the more detailed Blaha (2023a).

2.2 FRF and FRF Transformation Features

FRF features are best understood if one considers a dimension array with d_{dr} components and then map it to a vector also with d_{dr} components. We can then view the FRF transformations as occurring for these vectors in a higher dimension space, which we will call its *unification space*. The set of all FRF transformations of the unification space then can generate FRFs. An FRF transformation has $d_{dr} \cdot d_{dr}$ components.

The set of all FRF transformations also generate all transformations of symmetry groups when the dimension array, after a transformation, is mapped to irreducible symmetry group representation dimensions. Thus the designation *unification.*

The unification space for the dimension r space may be taken to be a Cosmos space. We assign it the dimension r′ where

$$r' = 2r + 4 \tag{2.1}$$

or

$$r' = 2r + 7 \tag{2.1a}$$

if one considers fermion strata and gambol fermions.

Based on the requirement that the unification space has a $d_{dr} \cdot d_{dr}$ components in its dimension array – the same number of components as the FRF transformations we see

$$d_{dr'} = d_{dr} \cdot d_{dr} = 2^{2r+8} = 2^{r'+4} \tag{2.2}$$

or alternately

$$d_{dr'} = d_{dr} \cdot d_{dr} = 2^{2r+14} = 2^{r'+7} \tag{2.2a}$$

if one considers fermion strata and gambol fermions. Fig. 2.1 shows the unification spaces for HyperCosmos spaces.

FRF transformations are generally not invertible. They can map a dimension array vector to one of many possible FRFs. The FRF transformations are many to one transformations. If a subsidiary condition is implemented then the augmented transformation can be made one to one.

Since the numeric value of dimensions is not relevant (Only their non-zero or zero values are relevant.) we can use simple numeric FRF transformations. Note that we are only dealing with Cosmos structure. Mass-energy considerations are not relevant.

THE HYPERCOSMOS HYPERUNIFICATION SPACES

Blaha Space Number $N = O_s$	Cayley-Dickson Number n	Cayley Number d_c	Dimension Array column length $d_{cd} = d_{cr}$	Dimension Array Size $d_{dN} = d_{dr}$	Space-time-Dimension r	Space-time r' Source of d_{dN} r'	HyperUnification Space Array d_{dN}'
0	10	1024	2048	2048^2	18	40	2^{44}
1	9	512	1024	1024^2	16	36	2^{40}
2	8	256	512	512^2	14	32	2^{36}
3	7	128	256	256^2	12	28	2^{32}
4	6	64	128	128^2	10	24	2^{28}
5	5	32	64	64^2	8	20	2^{24}
6	4	16	32	32^2	6	16	2^{20}
7	**3**	**8**	**16**	$\mathbf{16^2}$	**4**	**12**	$\mathbf{2^{16}}$
8	2	4	8	8^2	2	8	2^{12}
9	1	2	4	4^2	0	4	2^{8}

Figure 2.1. The HyperCosmos spaces spectrum related to HyperUnification spaces spectrum. Each dimension r is related to its HyperUnification space equvalent r'.

2.3 Transformations between a Dimension Array and a Single Dimension FRF

We now consider transformations between a space with a full dimension array (all non-zero components) and an FRF dimension array with only one non-zero dimension. One transformation maps the full array to the one dimension FRF. One can define an inverse transformation that maps a one dimension FRF to a full dimension array.

We will see later that the inverse transformations enable us to generate all the Cosmos space dimension arrays from one Penultimate Unification FRF. As a result we find a parallel to the generation of all Cosmos spaces from the single mathematical point (dimension) see in Blaha (2025i).

We will also see that the transition is entropically favored using our definition of Cosmos entropy. See chapters 1 and 3.

We now turn to identifying all unification spaces of Cosmos Theory. See Fig. 2.2. These spaces are discussed in detail in Blaha (2023a).

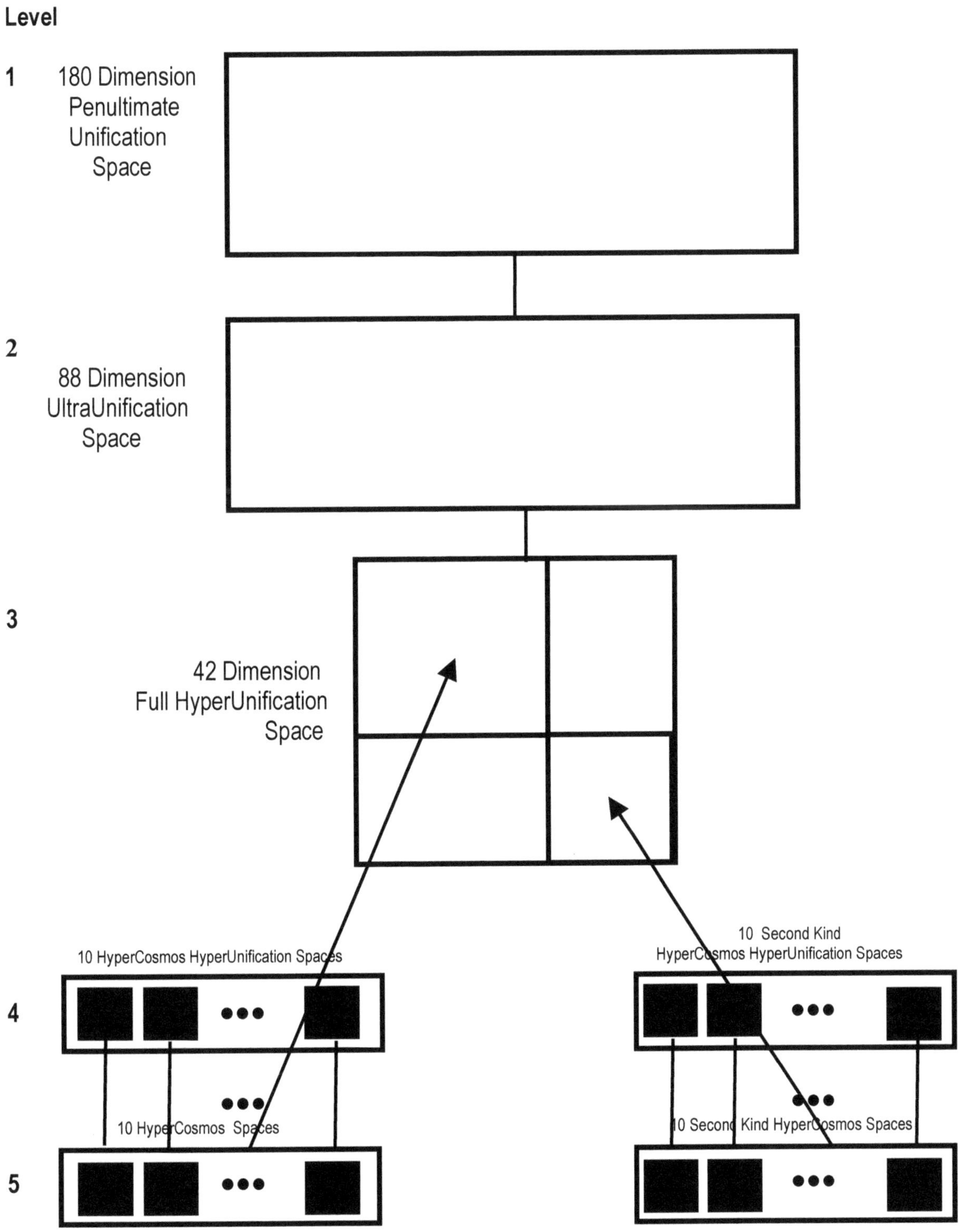

Figure 2.2. Five levels of the Cosmos spaces unification. They contain 43 spaces. From Blaha (2025g).

2.4 HyperUnification Spaces

Consider a Cosmos space of dimension r. It has a corresponding HyperUnification space with dimension[8] r' where the number of components in the r' dimension array column equals the total number of components in the dimension array d_{dN} of the r dimension array.

Each Cosmos space (Figs. 1.1–1.3) has a corresponding HyperUnification space. Fig. 2.1 lists unification space dimensions for the HyperCosmos spaces. Other sets of Cosmos spaces also have unification spaces. Fig. 2.2 lists the HyperUnification Cosmos spaces in level 4.

Each HyperCosmos space has a HyperUnification space. The FRF Transformations on the HyperUnification space dimension array have two important roles:

1. To support transformations to and from an FRF with a dimension array having ouly one non-zero dimension. Thus one can build a Cosmos space dimension array with a transformation from one non-zero dimension in the HyperUnification space's dimension array vector.

2. To support internal symmetry transformations on the r Cosmos space that implement symmetry group unification.

2.5 Full HyperUnification Space

The Full Unification space (level 3 in Fig. 2.2) combines the spaces of the twenty HyperCosmos and 2nd Kind HyperCosmos spaces of level 4. This space is a 42 dimension space. It has a dimension array with d_{dN_H} component dimensions where

$$d_{dN_H} = 8{,}388{,}608^2 = 7.03687\times10^{13} \tag{2.3}$$

It has d_{cN_H} columns and d_{cN_H} rows where

$$d_{cN_H} = 2^{42/2+2} = 2^{23} = 8{,}388{,}608 \tag{2.4}$$

In previous books we showed that the 42 dimension Full Unification space dimension array can be put in the block diagonal form displayed in Fig. 2.3 where

Subblock A has the dimension arrays of the 10 HyperCosmos spaces along its diagonal. Its column and row lengths are

$$d_{cS} = 5{,}592{,}400 \tag{2.5}$$

[8] In Cosmos Theory the dimension of a space equals the space-time dimension of the space by construction.

Subblock B has the dimension arrays of the 10 2nd Kind HyperCosmos spaces along its diagonal. It has d_{cB} columns and d_{cB} rows where

$$d_{cB} = 5{,}592{,}400/2 = 2{,}796{,}200 \qquad (2.6)$$

Subblock C is an 8 by 8 dimension array for a 2 dimension space. We assign this space to the role of providing 2 dimension rotations between corresponding HyperCosmos spaces and 2nd Kind HyperCosmos spaces. We have considered these possible rotations for spaces in Blaha (2025h). For example, its eq. 3.11 gives

$$S(\text{HyperCosmos, r, n}, \theta_n, \varphi_n) = e^{i\varphi_n}\cos(\theta_n)S_{\text{HyperCosmos, } n=n'=r+4} + e^{-i\varphi_n}\sin(\theta_n)S_{\text{2nd Kind HyperCosmos, } n\pm 1=n'=r+3\pm 2}$$

and eq. 3.13 gives

$$\Psi(r, \theta_r) = e^{i\varphi_r}\cos(\theta_r)\psi_H(r) + e^{-i\varphi_r}\sin(\theta_r)\psi_{H2}(r)$$

Section 5.5 of Blaha (2025h) provides an interpretation of their wave function superposition.

Consequently the column lengths (and row lengths) satisfy

$$d_{cN_H} = d_{cS} + d_{cB} + 8 \qquad (2.7)$$

where the length 8 is for the C subblock in Fig. 2.3 for a HyperCosmos space of dimension 2.

These calculations are for the case of the Cosmos spaces without strata fermions and without Gambol Fermions. They are described in detail in Blaha (2023a).

The FRF Transformations on the Full HyperUnification space dimension array have two important roles:

1. To support transformations to and from an FRF with a Full HyperUnification space dimension array having only one non-zero dimension. Thus one can build the 20 Cosmos space dimension arrays with a transformation from a one non-zero dimension Full HyperUnification space's dimension array vector.

2. To support internal symmetry transformations on the set of 20 Cosmos spaces that implement symmetry group unification across all the spaces.

2.6 UltraUnification Space

The UltraUnification space (level 2 in Fig. 2.2) is the unification space of the 42 dimension Full HyperUnification space. It is an 88 dimension Cosmos space: From eq. 2.1 we find the dimension

$$88 = 2\cdot 42 + 4 \qquad (2.8)$$

The FRF Transformations on the UltraUnification space dimension array have also two important roles:

1. To support transformations to and from an FRF with an UltraUnification space dimension array having ouly one non-zero dimension. Thus one can build the 42 Cosmos space dimension array with a transformation from a one non-zero dimension UltraUnification space's dimension array vector.

2. To support internal symmetry transformations on the 42 dimension Full Unification space that implements symmetry group unification for the entire space.

2.7 Penultimate Unification Space

The Penultimate Unification space (level 1 in Fig. 2.2) is the unification space of the 88 dimension UltraUnification space. It is a 180 dimension Cosmos space: From eq. 2.1 we find the dimension

$$180 = 2{\cdot}88 + 4 \qquad (2.9)$$

The FRF Transformations on the Penultimate Unification space dimension array also have two important roles:

1. To support transformations to and from an FRF with a Penultimate Unification space dimension array having only one non-zero dimension. Thus one can build the 88 Cosmos space dimension array with a transformation from a one non-zero dimension Penultimate Unification space's dimension array vector. This construction can cascade to build all 20 Cosmos space dimension arrays and thus the Cosmos structure.

2. To support internal symmetry transformations on the 88 dimension UltraUnification space that implements symmetry group unification for the entire space. Thus one achieves symmetry group unification throughout the Cosmos. The unification based on the Penultimate Unification space transformations completes the unification program.

2.8 The Unification Spaces

The 180 dimension Penultimate Unification space completes the Cosmos.[9] The Cosmos has four levels as shown in Fig. 2.2. The total number of Cosmos spaces is 43 spaces:

---

[9] The appearance of the dimensionless integers 42 and 88 in this Cosmos evokes the memory of the Eddington and Dirac (and others) conjectures on the appearance of 40 and 80 approximately in various combinations of universe parameters and physical constants. If some of their conjectures are valid they would be evidence for the author's Cosmos Theory. The dimensionless numbers 40 and 80 share the property of dimensionlessness with the dimensions of Cosmos Theory.

!0 HyperCosmos spaces + their 10 HyperUnification spaces + (2.10)
+ !0 Second Kind HyperCosmos spaces +
+ their 10 Second Kind HyperUnification spaces +
+ the 42 Dimension Full HyperUnification Space +
+ the 88 Dimension space
+ the 180 Dimension space

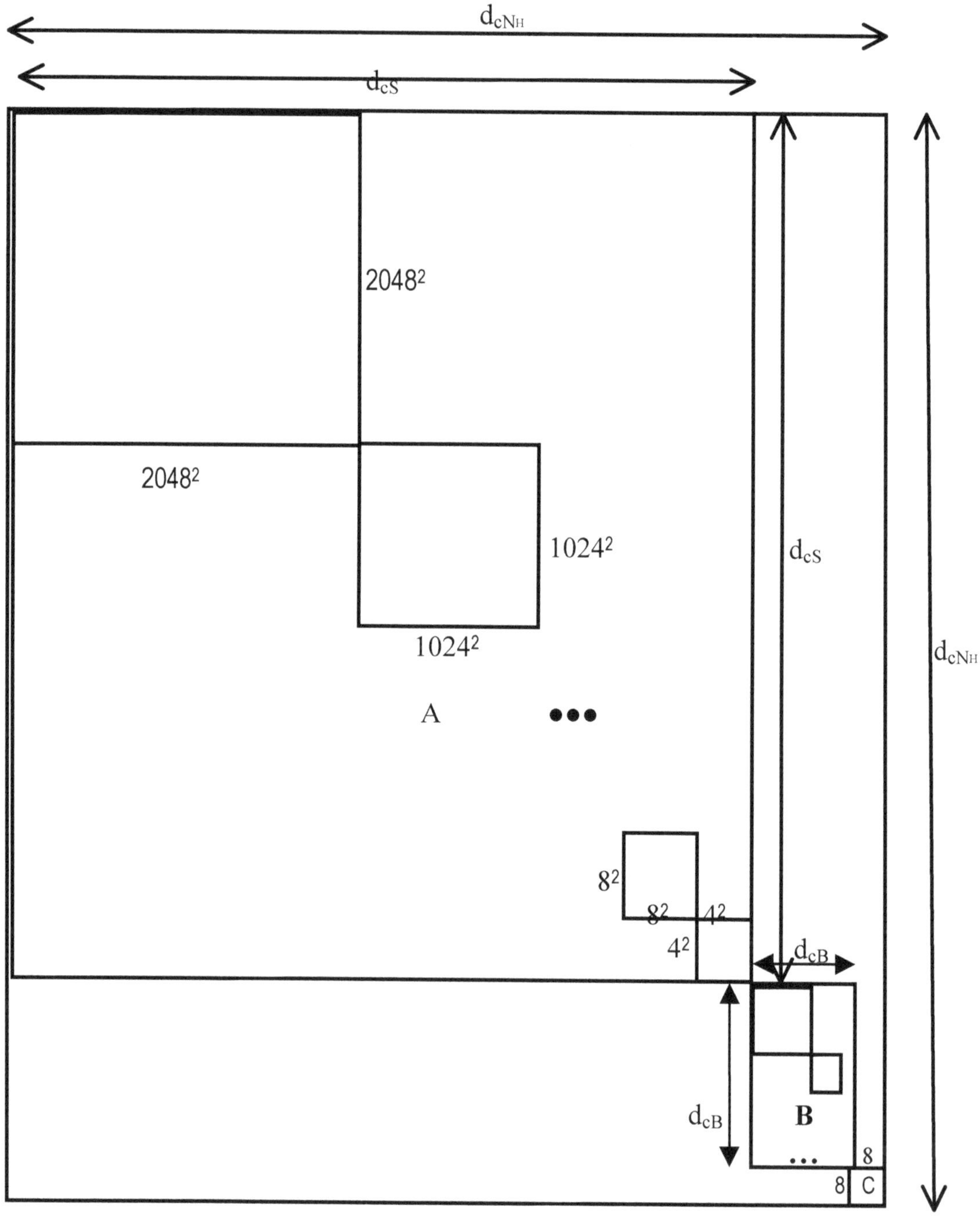

Figure 2.3. Form of the 42 dimension Full HyperUnification space dimension array. It is not drawn to scale.

3. Entropy from the Cosmos Penultimate Space

Time is a much overrated commodity. We trade time for money and call it work. When we measure time we use mechanical devices for clocks. We see that time measurements degenerate to be measurements of motion or of electromagnetic signals such as the decay of particles. What is it? We experience time as a flow and backflow of events. We experience time as passing quickly or slowly.

Now with the advent of Cosmos Theory we have a need to distinguish the eternal from the temporal. We have accomplished this differentiation by separating Cosmic structure from temporal events. We define Cosmos structure in terms of spaces with dimension arrays. They have no inherent matter or energy. They are unchangeable – not subject to forces or events. Immutable. They exist forever from the perspective of the time in any universe.

We have defined the structure of Cosmos Theory based on a series of happenings starting from a single mathematical point (or dimension) and proceeding to the generation of the Cosmos Fractal Curve and then to a Cosmos space governed by a Master Equation. Then we find a plethora of Cosmos spaces that we described in an earlier chapter and elsewhere. This development, which we necessarily describe sequentially to obtain a clear understanding of their origin, takes place instantaneously.

When? Do not ask that question. It is an eternal instantaneity – happening, as it is, outside of time. (We view time as only an artifact of the coordinate system of a universe.) What do we see if we could look at the structure of the Cosmos? We see an ever-present, unchanging Reality. From this perspective we are led to consider the introduction of mass and energy – mass-energy – within the framework of universes. We necessarily start with a Parent universe, which is co-eternal with Cosmos structure. It also exists forever. (See our earlier work justifying this view.) Within the Parent universe we suppose that child universes exist such as our universe. These universes have times. They evolve. They have interactions within them that cause internal change. The interactions (forces) within them are determined by maps from the internal structure of Cosmos dimension arrays.

This description of Cosmos Theory is supplemented by the view espoused in chapter 2 where we describe a 180 dimension Penultimate Unification space – a generated Cosmos space – that enables us to create the complete set of Cosmos spaces from one dimension. This creation mechanism parallels – *is* – the creation mechanism from a mathematical point. It offers the possibility of an entropic view of the generation of Cosmos structure.

Having that view we can see a necessity for Cosmos structure based on the increase of Cosmos entropy. We need not "force" the generation of the spectrum of Cosmos spaces from a mathematical point. It happens automatically based on entropic considerations. Then we have a clear Physically motivated sequence of Cosmos structure evolution. Again we note this evolution is instantaneous. Our discursive description of Cosmos evolution only clarifies the nature and features of Cosmos structure. The remaining part of Cosmos Theory – universes – will be addressed again in later chapters. It also is based on work in our previous books.

3.1 Entropy Based Generation of Cosmos Space Structure

There is a Cosmic entropy associated with the creation of the structure of Cosmos Theory: its spaces and dimension arrays. The "creation" is instantaneous and ever present since it has no associated time. Nevertheless we can view it as analogous to a sequence of Chemical reactions that starts with an initial mix of chemicals and proceeds through stages to a final mix of chemicals. Entropy increases as it passes through its stages resulting in a large entropy corresponding to a large set of chemical microstates.

We now consider the Cosmos case of an initial single non-zero dimension in the 180 dimension Penultimate Unification space. This dimension grows to a larger set of dimensions through a General-Relativistic-Internal-Symmetry transformation(s). This larger set of dimensions then becomes a yet larger set through another General-Relativistic-Internal-Symmetry transformation(s) in the 88 dimension UltraUnification space. And so on until the set of 20 Cosmos spaces is reached. The stages are:

180 Dimension Penultimate Unification Space
88 Dimension UltraUnification Space
42 Dimension Full Unification Space
The 20 HyperCosmos and 2nd Kind HyperCosmos Spaces

The crucial quantities are the row lengths of the spaces which are determined by

$$d_{cr} = 2^{r/2+2} \tag{3.1}$$

where r is the space-time dimension. The starting point of the construction is one non-zero dimension in the dimension array of the 180 Dimension Penultimate Unification Space It gives the initial entropy:

$$S_e = k_{Cosmos} \ln_2(1) = 0 \tag{3.2}$$

The one non-zero dimension must grow through a General-Relativistic-Internal-Symmetry transformation(s) to fill a column of non-zero dimensions in the Penultimate Unification dimension array as in section 1.3. This column has 2^{92} dimensions (92 = 184/2). Consequently the entropy of the filled column is

$$S_{e180} = \ln_2(2^{92}) = 92 \tag{3.3}$$

Then the process continues as displayed in Fig. 2.2 descending level by level. The dimension arrays of the spaces below the 180 space are filled.

3.2 The Process of Generating the Cosmos Spaces' Dimension Arrays

The process goes through a number of phases with the generation processes of the set of spaces leading to the filling of the 20 Cosmos spaces. Each step in the process is due to a General-Relativistic-Symmetry-Group transformation(s).

1. The filled 180 Penultimate Unification space dimension array column is the dimension array of the 88 dimension UltraUnification dimension array.
   Change in number of filled dimensions = $2^{92} = 2^{180/2+2} = 2^{88+4}$

2. The "top" vector in the filled 88 dimension UltraUnification space dimension array is the dimension array of the 42 dimension Full Unification space.
   Change in number of filled dimensions = $2^{46} = 2^{88/2+2} = 2^{42+4}$

3. The "top" vector in the filled 42 dimension Full Unification space dimension array is the vector composed of all 20 dimension *arrays* of the 20 Cosmos spaces.
   Change in number[10] of filled dimensions $\cong 5 \cdot 2^{43} = 4.4 \cdot 10^{13}$ = sum of 20 Cosmos dimension array sizes (as in Figs, 2,2 and 2.3)

3.3 The Entropy of the Generation of the Cosmos Spaces from One Non-zero Dimension in Penultimate Unification Space

The three subprocesses above contribute to the total entropy of the generation. We use the method of section 1.3 where the entropy is based on the number of non-zero dimensions n:

$$S_e = \ln_2(n) \tag{3.4}$$

In calculating the total entropy we use the minimal[11] value of the entropy for each subprocess separately assuming the subprocess starts with zero entropy. It changes by the amount of the change in dimensions.

We find

$$S_{e180} = \ln_2(2^{92}) = 92$$
$$S_{e88} = \ln_2(2^{46}) = 46$$
$$S_{e42} = \ln_2(5 \cdot 2^{43}) = 45.3$$

The total change in entropy is

---

[10] This value follows from eqs. 2.4 and 2.5.

[11] The entropy values in subprocesses 1 and 2 above are for the (*minimal*) filling of one row/column of the dimension arrays of the 180, 88, and 42 dimension spaces. Thus the stated entropy amounts are less than they would be if all rows of their dimension arrays were made non-zero. The goal was to find the minimal total entropy for the process, which determines the 20 Cosmos spaces' dimension arrays starting from one non-zero dimension in the 180 dimension space's dimension array. Note: the "top" vector of the 42 dimension space contains *all* of the 20 spaces dimension arrays in a linearized vector form.

$$S_{eTotal} = 92 + 46 + 45.3 = 183.3$$

The entropy increases as the original one non-zero dimension blossoms into the very large number of dimensions of the Cosmos spaces.

We conclude that the process of generating the set of 20 Cosmos spaces dimension arrays has a Cosmos entropy of 183.3. This inexorable process is instantaneous and automatic. It takes a single microstate and generates a high entropy state with numerous microstates.

3.4 Interpretation of General-Relativistic-Symmetry-Group Transformation

This type of transformation was discussed in detail in our 1970s papers.[12] They are a form of relativistic transformation between coordinate systems that includes symmetry group transformations. Those papers showed these transformations occur in the operators of quantum fermion field Fourier expansions. In the present case this transformation implements a transformation of the Cosmic entropy due to the equivalent of a change of coordinate system. Chapter 2 describes it in some detail.

The present case is analogous to the transformation of a fermion state to its form in the accelerating coordinate system. The fermion, in the accelerating coordinate system, appears to be a set of fermions. Thus one fermion in the unaccelerated coordinate system "becomes" many fermions in the accelerating coordinate system. The number of fermions changes just as the number of dimensions change in the entropy case discussed above. Here the change is a change in Cosmic entropy due to a change in the number of non-zero dimensions.

Entropy inexorably guides the dimension generation process without outside intervention.

[12] S. Blaha, Phys. Rev. **D17**, 994 (1978); S. Blaha, "The Local Definition of Asymptotic Particle States", IL Nuovo Cimento **49A**, 35 (1979).

4. Parent Universe(s)

A Parent universe necessarily appears in Cosmos Theory based on a Consistency Condition that balances the Casimir force of compression against the thermodynamic pressure of the particles within the Parent universe. The number of particles per unit volume, determined by the dimension array size, is fixed and specifies the dimension of the Parent universe. The Parent universe must be instantaneous with the other components of Cosmos Theory. Thus it also must exist forever.

We thus have a complete set of Cosmos Theory components. They are indicated in Fig. 4.1. As part of this view we note that the Parent universe, which must have a quantum realization, must be a quantum field in Cosmos Space, the two plus one dimension space generated by the Cosmos Fractal Curve *simultaneously*. The Parent universe quantum field is now identified as the fourth component of Cosmos Theory. See Fig. 4.1.

4.1 Parent Universe Consistency Condition

The number of Cosmos spaces is infinite in principle. Their dimensions range from $-\infty$ to $+\infty$. Some are listed in Figs. 1.1 – 1.3. A Physical requirement for a "long-lived" viable space is that it does not collapse to nothing due to a Casimir vacuum-based force. The force should be balanced (or exceeded) by the internal outward thermodynamic pressure of the contents of the universe.

We applied this Consistency condition to determine the upper limit on the set of Physical Cosmos spaces. The limiting dimension was found to be 18. Thus we have a *Physical* set of 10 HyperCosmos spaces and 10 2nd Kind HyperCosmos spaces with dimensions being even numbered integers where the dimensions range from 0 through 18. We define the Parent universe to have the 18th dimension. It is the least dimension universe where the universe does not collapse from the Casimir force. Blaha (2025i) provides the details of the Consistency Condition calculation.[13]

For a sphere (actually a HyperSphere) of radius a we found the thermodynamic outward pressure in an r dimension space-time:

$$P = N\,\Gamma((r-1)/2+1)kT/(\pi^{(r-1)/2}a^{(r-1)}) \qquad (4.1)$$

where $\Gamma(n)$ is the factorial function. The number of fermions N is given by eq. 6.14.

Turning to the vacuum energy and Casimir force we found the energy deficit per unit volume:

$$U \cong \pi^{(r-1)/2}a^{-r}(r-1)/[r\Gamma((r-1)/2+1)] \qquad (4.2)$$

[13] Equations numbered 6.xxx in the following are from Blaha (2025i). We did not change these numbers for the convenience of the reader who might wish to refer to that volume.

Multiplying by the volume of the bubble we find a total bubble energy:

$$U_{tot} \approx V_{r-1}\, U \tag{4.3}$$

$$\approx \pi^{r-1} a^{-1}[(r-1)/r]/\Gamma((r-1)/2+1)^2 \tag{4.4}$$

Then

$$\partial U_{tot}/\partial a \cong -\pi^{r-1} a^{-2}[(r-1)/r]/\Gamma((r-1)/2+1)^2 \tag{4.5}$$

A bubble surface of radius a has the area

$$S_{r-2} = 2\pi^{(r-1)/2} a^{r-2}/\Gamma((r-1)/2) \tag{4.6}$$

The Casimir force on the surface is therefore

$$\begin{aligned} C &= (S_{r-2})^{-1}\, \partial U_{tot}/\partial a \\ &= -\tfrac{1}{2}\pi^{(r-1)/2} a^{-r}[(r-1)/r]\,\Gamma((r-1)/2)/\Gamma((r-1)/2+1)^2 \\ &= -\pi^{(r-1)/2} a^{-r}/[r\Gamma((r-1)/2+1)] \end{aligned} \tag{4.7}$$

The Consistency Condition is

$$H = P - C = 0 \tag{4.8}$$

It determined the maximum dimension to be 18, which is the lowest dimension that supports a non-collapsing universe. The result is the Parent universe dimension and the set of *Physical* Cosmos spaces.

The Parent universe coexists with the other components of Cosmos Theory. Although we view it as sequential in these descriptions for the purpose of understanding its origin, it also exists forever. The Parent universe, whose quantum field is a solution of the Master Equation, appears in Fig. 4.1 as one of the four eternal components of the Cosmos. See section 4.3 below.

4.2 Parent Universe Energy

The derivation of the Casimir energy and force above gives a "missing" bubble vacuum energy per unit volume

$$U_{bubble} = \int_{1/a}^{\infty} d^{r-1}k\, (k^2 + m^2)^{1/2} \tag{4.9}$$

The missing vacuum energy constitutes a positive energy for the Parent universe, which is, after all, the bubble. This energy diverges.[14] Consequently the energy of the Parent universe is a non-zero, but unspecified quantity based on the energy distortion of the universe. The derivative of the vacuum energy eliminates the divergent constant and is used in the Consistency Condition.

[14] The Casimir force being the derivative of the energy does no include the divergent constant energy.

The Parent universe energy can also be estimated from its average temperature and the number of fermions if one assumes the equipartition of energy. It is the energy of motion of the particles in the Parent universe since all particles are initially massless and did not exist prior to mass generating interactions. We take the energy E_{Parent} to be

$$E_{Parent} = NkT$$

where N is the total number of fermions. Using

$$N_{Cosmos} = \cos(\theta)\, 2^{r+7} + \sin(\theta)\, 2^{r+6} \quad (6.14)$$

with

$$\cos(\theta) = 1$$

from Blaha(2023i) where we found

$$akT = 2.85 \times 10^{-12} \quad (6.34)$$

Thus

$$E_{Parent} = NkT = 2.85 \times 10^{-12}\, N/a$$

From

$$N_{CosmosMAX} = 5 \cdot 2^{r+5} \cos(\theta) = 5 \cdot 2^{r+5} \quad (6.33)$$

we obtain

$$E_{Parent} = 1.21 \times 10^{-4}/a$$

for r = 18. Since a is presumably, extremely small, the energy of the Parent universe must be extremely large.

4.2.1 Energies of the Universes of Different Spaces

The energy[15] E = number of dimension array fermions of the Parent universe appears in its quantum field in eq. 4.22 below. Its conjugate variable is t. We determined t from the running length of the Cosmos Fractal Curve measured in construction steps n. Thus we may[16] take $t = 2^n$. We have also identified each construction step n with a Cosmos space with dimension array size of 2^{r+4} where r is the space-time dimension and $n = r + 4$.[17]

The known energy – time uncertainty relation is

$$[E, t] = i\hbar$$

For Cosmos spaces we suggest

---

[15] This energy is not the E appearing in the evaluation of the Master Equation solutions.

[16] The line length in step n is 2^n. We will generalize n to be continuous in this section.

[17] For 2nd Kind HyperCosmos spaces we set $n = r + 3$ and set the array size to 2^{r+3}. Introducing fermion strata and gambol fermions would change n to r+7 and r+6 for HyperCosmos and 2nd Kind HyperCosmos spaces respectively.

$$[E, 2^n] = i\hbar \tag{4.10}$$

based on the above observations. Consequently, noting

$$\Delta E \Delta(2^n) \geq \hbar \tag{4.10a}$$

we find the conjugate variables have an inverse relation from the viewpoint of Fourier transforms:

$$E \sim 1/2^n \tag{4.10b}$$

Since n = 22 for the Parent universe, one may estimate the energy of an n′ = r + 4 Cosmos space *child* universe to be a fraction of the Parent universe energy

$$2^{n'}/2^{22} \tag{4.10c}$$

where n′ < 22. In general, for $n' < n_{Parent}$

$$E_{n'}/E_{Parent} \approx 2^{n'}/2^{Parent} = n'/n_{Parent} \tag{4.10d}$$

For our universe where n′ = 8 we find the very small ratio 2^{-14} = 1/16384. *There is room for up to 16384 child universes of the type of our universe within the Parent universe.*

Using the relation seen above

$$E = NkT \tag{4.10e}$$

and comparing to eq. 4.10d, we see the initial temperature of the child universe equals the temperature of the Parent universe:

$$kT = kT_{Parent} \tag{4.10f}$$

We may attribute this equality to a common origin in a fractal: the Parent universe in the Cosmos Fractal Curve and the child universe in a HyperSphere fractal curve (discussed later).

Later we will consider the generation of child universes within the Parent universe. (We do not exclude the possibility of multiple, probably non-interacting, Parent universes with each having its own hierarchy of child universes.)

4.3 Parent Universe Quantum Field Solution

We now find the quantum field for the Parent universe in the Cosmos Space at radial coordinate u and time t. The initial form of the quantum field does not explicitly embody the Parent universe space's dimension array. In section 4.4 we will expand the dimensions for the quantum field Extended Master Equation to include the dimensions of the dimension array including an explicit internal Parent universe space-time coordinate system.

The Cosmos Fractal Curve defines an infinite two dimension space, to which we add a time coordinate equal to the running length of the Fractal Cosmos Curve as it was constructed. Thus there are 2+1 dimensions. We call this space the Cosmos Space. In the space we have defined a Lagrangian for the Master Equation.[18] The Master Equation generates an eigenvalue spectrum and a set of wave function solutions.

The Lagrangian density is

$$\mathcal{L} = ½(\partial^\mu \psi(z)\partial_\mu\psi(z)/m \; - V(z)\psi(z)\psi(z))$$

where z is the 3-vector (t, x, y), $\partial^\mu\psi(z)$ is the covariant derivative, a (non-mass) constant m is introduced for use later, and V(z) is a scaling potential

$$V(z) = \lambda/[m^2(x^2 + y^2)] \tag{4.11}$$

with λ a constant, From it we derived the Master Equation:

$$[(\partial^2/\partial x^2 + \partial^2/\partial y^2)/m + \lambda/m^2r^2 + C]\psi(x, y) = 0 \tag{4.12}$$

where $u = (x^2 + y^2)^{½}$ after letting

$$\psi(t, x, y) = e^{iEt}\psi(x, y) \tag{4.13}$$

with $mC = E^2$.

4.3.1 Wave Function Solutions

The Master Equation becomes

$$(\partial^2/\partial u^2 + 1/u\; \partial/\partial u + \partial^2/\partial\varphi^2 + \lambda/mu^2 + mC)\psi = 0 \tag{4.14}$$
$$(\partial^2/\partial\varphi^2)\psi = -k^2\psi$$

We define

$$\lambda_{n\varphi} = (\lambda - k^2)/m \tag{4.15}$$

We now define

$$\lambda_{n\varphi} = n_\varphi{}^2/m \tag{4.16}$$

The form of the complete wave function solution of the Master Equation is

$$\psi_{E,\, k,\, n\varphi,\, n}(u, t) = e^{iEt}\; \exp^{ik\varphi}\psi_{n\varphi n}(u) \tag{4.17}$$

where we use polar coordinates (r, φ). The ψ(u) radial factor solution is

$$\psi_{n\varphi n}(u) \approx B\cos((\lambda_{n\varphi} - ¼)^{½}\ln((-mC_{n\varphi n})^{½}u) + b) \tag{4.18}$$

---

[18] This procedure is discussed in detail in Blaha (2025h).

where $c_{n\varphi}$, b, $C_{n\varphi n}$, and B are constants. The integer constant n is the radial quantum number. The integer constant n_φ is an extension of the angular quantum number.

The quantum field is expressed in positive and negative "frequency" parts using the identity:

$$\cos(\ln(2^n u)^\lambda) = (\exp(i\ln(2^n u)^\lambda) + \exp(-i\ln(2^n u)^\lambda))/2 \quad (4.19)$$
$$= (\exp(i\lambda \ln(2^n u)) + \exp(-i\lambda \ln(2^n u)))$$
$$= (e^{-i\lambda X - i\lambda P} + e^{i\lambda X + i\lambda P})$$

based on eq. 4.18 where

$$X = \ln u \quad (4.20)$$
$$P = \ln 2^n$$
$$E = 2^n$$

for a coordinate parameter u and constant λ.

Treating the quantum field as spin 0 we find the general form of the Fourier expansion of the solution of *eq. 4.14*:

$$\psi_{n\varphi}(\varphi, u, t) = \sum_n \int dk\, N(k, n)\{a_{n\varphi}(k, n)\exp(-iEt)e^{-ik\varphi}(2^n u)^{-i\lambda} +$$
$$+ a_{n\varphi}^\dagger(k, n) \exp(iEt)e^{ik\varphi}(2^n u)^{i\lambda}\} \quad (4.21)$$

and

$$\psi_{n\varphi}(\varphi, u, t) = \sum_n \int dk\, N(k, n)\{a_{n\varphi}(k, n)\, e^{-iEt - ik\varphi - i\lambda X - i\lambda P} +$$
$$+ a_{n\varphi}^\dagger(k, n)\, e^{iEt + ik\varphi + i\lambda X + i\lambda P}\} \quad (4.22)$$

as seen in Blaha (2025i). This scalar quantum field $\psi_{n\varphi}(\varphi, u, t)$ is the quantum field of the Parent universe in the Cosmos Space at polar coordinate positions u and φ.

4.4 Extended Master Equation Dimensions and its Parent Quantum Field

We now introduce a generalization of the Master Equation that has a larger set of coordinates beyond the three Cosmos Space coordinates. Each dimension in the 18 dimension Parent universe dimension array with 2^{22} dimensions has a corresponding coordinate in the Extended Master Equation. Of these 2^{22} dimensions, we reserve 18 dimensions as space-time dimensions.

4.4.1 Justification of the New Coordinates

There are several reasons for introducing new coordinates into the Master Equation:

1. The values of the quantized energy E's specify the size of dimension arrays. Thus each energy can be viewed as degenerate – as representing a degenerate set of component dimensions. One can use the new coordinates to distinguish between the components in the dimension array. This is similar to the case of the

hydrogen atom where the angles factor $L^2 = l(l+1)$ represents the $2l+1$ degenerate angular momentum states.

2. It unifies all interactions in the Parent universe including space-time. The set of dimension array components can be viewed as a vector in the $SU(2^{22})$ irreducible representation. Then we can view the components arranged in rectangular form.

3. The three dimension Cosmos Space was generated from the Cosmos Fractal Curve. The 2^{22} component space may also be viewed as generated from 2^{21} fractal curves – that generate a HyperSphere in 2^{22} dimensions. Each of the 2^{21} fractals give two dimensions to the HyperSphere.[19] Dimension arrays have a fractal form. Each of the HyperSphere's 2^{22} dimensions has a corresponding dimension within the dimension array. See Fig. 4.1. The HyperSphere Hausdorff fractal dimension D is

$$D = \ln_2(2^{22}) = 22 \tag{4.23}$$

The new dimensions unite with the three Cosmos Space dimensions to form the Extended Master Equation.

4. The $SU(2^{22})$ group of the Parent universe coordinates facilitates the mappings of other spaces' dimension arrays to the symmetry groups of their universes.

The form of the Extended Master Equation is:

$$[(\partial^2/\partial x^2 + \partial^2/\partial y^2)/m + \lambda/m^2r^2 - \partial^2/\partial t^2 + \sum_{i=1}^{2^{22}} \partial^2/\partial x_i^{\,2}]\, \psi(x, y, t, x_1, x_2, \ldots, x_{2^{22}}) = 0 \tag{4.24}$$

based on eq. 4.12. Letting

$$\psi(x, y, t, x_1, x_2, \ldots, x_{2^{22}}) = e^{iEt}\, \psi(x, y, x_1, x_2, \ldots, x_{2^{22}})$$

leads to

$$[(\partial^2/\partial x^2 + \partial^2/\partial y^2)/m + \lambda/m^2r^2 + E^2 + \sum_{i=1}^{2^{22}} \partial^2/\partial x_i^{\,2}]\, \psi(x, y, x_1, x_2, \ldots, x_{2^{22}}) = 0 \tag{4.25}$$

Separating the equation with

$$\psi(x, y, x_1, x_2, \ldots, x_{2^{22}}) = \psi(x, y)\, \chi(x_1, x_2, \ldots, x_{2^{22}}) \tag{4.26}$$

and using

---

[19] See Blaha (2025i). Fig. 4.1 shows the Cosmos spaces' fractal curves in its second column. Note that a 221 dimension HyperSphere

We find

$$\sum_{i=1}^{2^{22}} \partial^2/\partial x_i^2 \, \chi = \alpha \, \chi$$

$$[(\partial^2/\partial x^2 + \partial^2/\partial y^2)/m + \lambda/m^2r^2 + (E^2 + \alpha)] \, \psi(x, y) = 0 \qquad (4.27)$$

or

$$[(\partial^2/\partial x^2 + \partial^2/\partial y^2)/m + \lambda/m^2r^2 + C'] \, \psi(x, y) = 0 \qquad (4.28)$$

where

$$C' = E^2 + \alpha \qquad (4.29)$$

thus recovering the form of eq. 4.12 with a "new" value C′. The form of the Master Equation solution remains:

$$\psi_{n\varphi n}(u) \approx B\cos((\lambda_{n\varphi} - 1/4)^{1/2}\ln((-mC'_{n\varphi n})^{1/2}u) + b) \qquad (4.30)$$

The SU(2^{22}) symmetry is unbroken. We can summarize the effects of the new dimensions as a shift in the energy E in the Master Equation due to the introduction of an additional set of coordinates for the Parent universe:

$$m(C' + \alpha) = E^2 \qquad (4.31)$$

With $C' = 2^n$ we find

$$E = m^{1/2}(2^n - \alpha)^{1/2} \qquad (4.32)$$

The value of C′ is the size of the dimension array. Its dimensions are place holders. Their numeric value is unimportant.

We will use the same procedure for child universes with a $2^n = 2^{r+4}$ dimension fractal HyperSphere and an r dimension space-time.

4.4.2 The Construction of the New Coordinates

In Blaha (2025i) we showed the process for generating the three dimensions of the Cosmos Space from the construction of the Cosmos Fractal Curve. We now similarly generate the 2^{22} new dimensions seen above from HyperSphere (HyperCube) fractal curves for the Parent universe. Later we will apply HyperSphere construction to generate child universe coordinates. *The construction of the new coordinates (dimensions) from 2^{21} fractal curves closely parallels the construction of the two Cosmos Space coordinates from a single fractal – the Cosmos Fractal Curve.*

We begin by defining a construction process that starts with one dimension line segments as discussed for the Cosmos Fractal Curve construction and grows to generate a 2^{22} dimension unit HyperSphere. This HyperSphere is generated from 2^{21} fractal curves. Each fractal curve generates a *two* dimension unit square which is made continuous and then scaled upwards to a two dimension space. The combined set of two dimension spaces gives a $2\cdot 2^{21} = 2^{22}$ dimension space. Each dimension of this space corresponds to a dimension in the Parent space and Parent universe dimension array. It is used in the Extended Master Equation. See eq. 4.24.

The initial HyperSphere construction steps use piecewise linear line segments with lengths growing by factors of

$$N = 2^n \tag{4.33}$$

where n is an integer. The generated HyperSphere fractal (and Physical) dimension is[20]

$$D_{HyperSphere} = \ln(2^d)/\ln(\varepsilon) = 2^{22} \tag{4.34}$$

where $d = 2^{22}$ for the Parent universe.

With our previous choice of construction steps, in prior chapters, we obtained a sequence of values of piecewise linear line copies of length 2^n to create one fractal curve. We obtain a D = 2 dimension fractal space-filling unit square, the Cosmos Fractal Curve,[21] that becomes the two dimension part of the Cosmos Master Equation space (neglecting the time dimension).

Similarly, we now construct a d dimension HyperSphere by subdividing each piecewise linear line copy of length 2^n into d/2 piecewise linear lengths. Each of these $2^{n+1}/d$ piecewise linear lengths can be scaled upward to become the piecewise linear lengths of a fractal curve. As a result we now have d/2 fractal curves.

Each of the d/2 fractal curves then specifies a dimension of a space-filled unit HyperSphere of dimension d. We now have a d dimension HyperSphere (HyperCube) that can be scaled to form a $d = 2^{22}$ dimension space since each fractal generates a two dimension space. The HyperSphere generated space, and its coordinates, can be extended to be continuous, and scaled to be of infinite extent, to support a space with rotations of its coordinates.[22]

Then these dimensions are used to generate the additional Master Equation coordinates. The combined set of three Cosmos coordinates and the HyperSphere generated coordinates then become the coordinates of the Extended Master Equation described above.

The HyperSphere generated coordinates also play the role of internal dimensions of the universe (the Parent universe in this case). They define the dimension array of the

---

20 Note: We use the Hausdorff fractal dimension expression:

$$D = \ln(N)/\ln(\varepsilon)$$

with N = the number of copies of the shape formed from each copy of the previous iteration of the curve under construction, and $\varepsilon = 2$ = the number of copies are scaled downward from the previous step. (Each step length is twice the previous step length.) In the present case there are 2^d copies of the shape that is formed from each copy of the previous iteration of the curve under construction.

21 Without the part that generates the initial unit line segment.

22 The map discussed here to a HyperSphere is the same map previously described by the author as an inverse map of a one dimension fractal to a Z-order Lebesgue d dimension space filling curve (Morton space-filling curve) and thence to a d dimension universe (space).

Parent space and universe. Each dimension in the dimension array has a corresponding coordinate.

Within that dimension array we define an r = 18 dimension *space-time* set of coordinates. Starting with the relation for the size of the dimension array:

$$D_{\text{dimension array}} = 2^{r+4} \tag{4.35}$$

we use eq. 4.34 to obtain

$$D_{\text{dimension array}} = D_{\text{HyperSphere}} = \ln(2^d)/\ln(\varepsilon) = 2^{22} \tag{4.36}$$

The Parent universe is defined by the dimension array of size

$$D_{\text{dimension array}} = 2^{22} \tag{4.37}$$

The space-time dimension of the Parent universe is

$$r = 18 \tag{4.38}$$

consistent with eq. 4.35.

The other $2^{22} - 18$ components of the Parent's dimension array are assigned to internal symmetries.

4.4.3 Symmetry Group of New Coordinates

The symmetry group of the new coordinates is $SU(2^{22})$. The quantum field solution transforms as an $SU(2^{22})$ irreducible representation. Since there are no interactions at this stage of the Extended Master Equation discussion the group is unbroken. The Parent universe mass-energy interactions that may appear can cause child universes to be created and interact. The application of $SU(2^{22})$ transformations to the $SU(2^{22})$ coordinates provides the symmetry of the new Lagrangian.

When child universes are created their symmetry groups are subgroups of the Parent universe symmetry group. Within the child universes interactions happen as part of their evolution. We discuss this later.

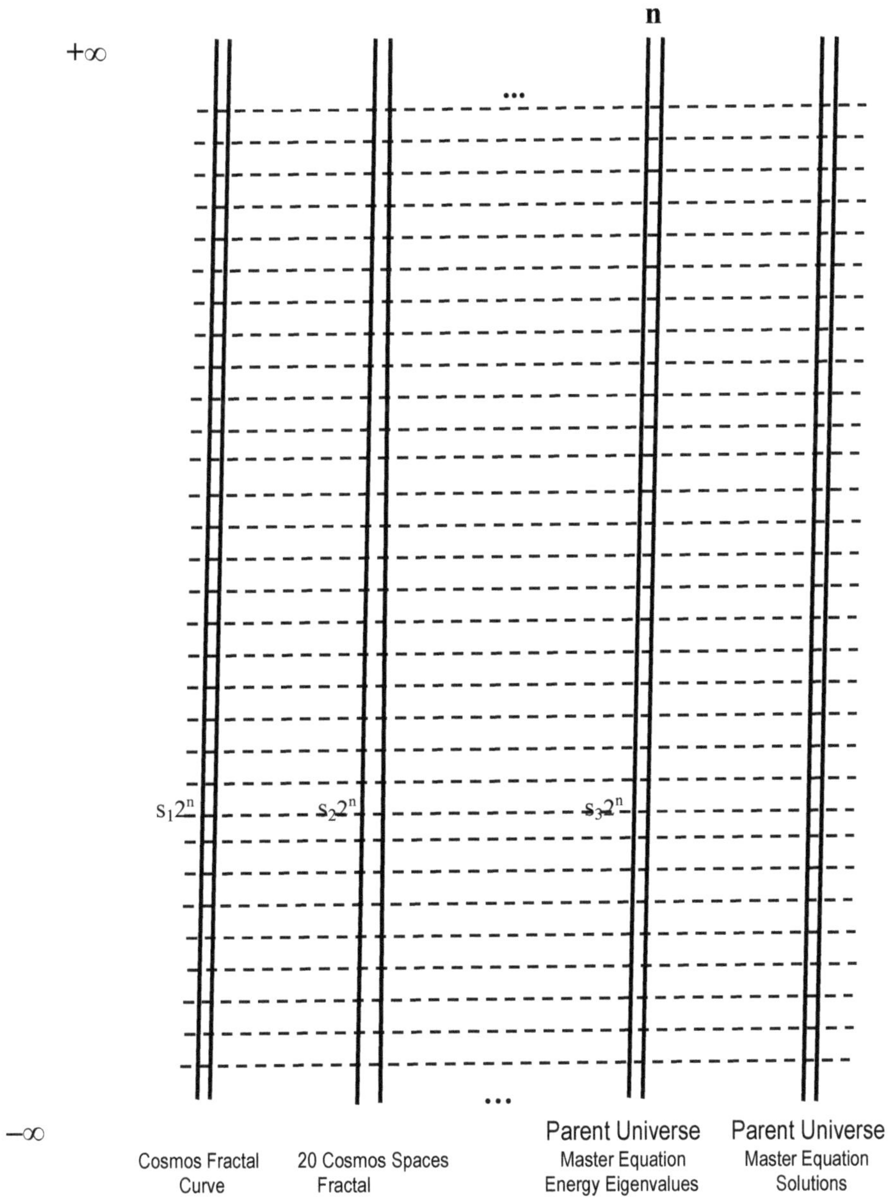

Figure 4.1. The four components of Cosmos Theory. The existence of HyperCosmos and 2nd Kind HyperCosmos spaces causes pairings. The constants s_1, s_2, and s_3 are overall factors for each sequence. The fourth component is the set of Parent universe quantum wave functions $\psi_{n\varphi n}(u)$ where u is the radial coordinate in Cosmos Space.

5. Child Universes and HyperSphere Space-Times

The beginning stages of our universe are understood to some extent. There are several theories that describe its evolution. However the true *beginning* of our universe remains clouded. Where did its mass-energy come from? Where were its coordinates determined? Why did it apparently begin as a point?

There have been a number of attempts to answer these questions including the author's books suggesting a Quantum Field Theory of universe interactions and a Big Bang theory.

One can view the creation of a universe as the result of vacuum effects within the Parent universe or as the result of interuniverse interactions within some Quantum Field Theory.

In this chapter we will consider the generation of a universe as a fractal process along the lines that we considered for the generation of the Parent universe. The Parent universe arose (or existed) with no other universes present and not within a universe with a dynamic vacuum.

So we have grounds to examine the issue of a universe coming into existence independent of considerations of other universes, interactions and a dynamic vacuum. Instead we treat the beginning of a universe as a fractal starting from a single point – perhaps possessing an energy.

5.1 Universe Dimension Arrays and Space-times

We will view the generation of a universe as beginning with a point that evolves into a set of fractal curves that generate an associated set of dimensions placed in a dimension array. HyperCosmos spaces' dimension arrays have[23]

$$2^{r+4} \tag{5.1}$$

components where r is the dimension of the space and *also* the dimension of the set of space-time coordinates of the universe. If one considers the enhanced Cosmos fundamental fermion spectra that include fermion strata and Gambol Fermions then the size of the dimension array is 2^{r+7}. We will consider the case of eq. 5.1 in this chapter. The case of 2nd Kind Cosmos Spaces and universes is similar (section 5.3).

5.2 Universe HyperSpheres

Consider a universe of r dimensions (and r space-time dimensions). We will construct a $d = 2^{r+4}$ dimension HyperSphere from d/2 fractals. Each fractal generator of

---

[23] 2nd Kind HyperCosmos spaces' dimension arrays have 2^{r+3} components. They have 2^{r+6} if they support fermion strata and Gambol Fermions.

the HyperSphere contributes two dimensions in the initial form of its unit square. Consequently the size of the universe's dimension array will be d.

In a manner similar to the HyperSphere construction for the Parent universe we set

$$d = 2^{r+4} \tag{5.2}$$

with the resulting dimension array size

$$d_{dr} = d = 2^{r+4} \tag{5.3}$$

where d_{dr} values may be seen in Figs. 1.1 – 1.3.

We construct the d dimension HyperSphere by subdividing each Hilbert Curve (Cosmos Fractal Curve) piecewise linear line copy of length 2^n into d/2 piecewise linear lengths. Each of these $2^{n+1}/d$ piecewise linear lengths can be scaled upward to become the piecewise linear lengths of a fractal curve. As a result we now have d/2 fractal curves.

Each of the d/2 fractal curves then specifies two (a unit square extended to a two dimension space) dimensions. The total number of dimensions generated by the HyperSphere is d. The HyperSphere is generated by d/2 fractal curves.[24]

We now have a d dimension unit HyperSphere (HyperCube) that can be scaled to form a d dimension array's space of 2^{r+4} dimensions since each fractal generates a two dimension space. The HyperSphere generated space, and its coordinates, can be extended to be continuous, and scaled to be of infinite extent, to support a space with rotations of its coordinates.

The universe HyperSphere generated coordinates play the role of internal dimensions of the universe (the child universe in this case). They define the dimension array of the child space and universe. Each dimension in the dimension array has a corresponding HyperSphere coordinate.

Within that dimension array we define an r dimension *space-time* set of coordinates using the relations

$$D_{\text{dimension array}} = d = 2^{r+4} \tag{5.4}$$

where

$$D_{\text{dimension array}} = D_{\text{HyperSphere}} = \ln(2^d)/\ln(\varepsilon) = 2^{r+4} \tag{5.5}$$

to obtain the space-time dimension of the child universe

$$r = \ln_2(d) - 4 = \ln_2(D_{\text{dimension array}}/16) \tag{5.6}$$

[24] The map discussed here to a HyperSphere is the same map previously described by the author as an inverse map of a one dimension fractal to a Z-order Lebesgue d dimension space filling curve (Morton space-filling curve) and thence to a d dimension universe (space).

The other $2^{r+4} - r$ components of the child universe's dimension array are assigned to internal symmetries.

In summary, a d dimension HyperSphere generates a d size dimension array and a d set of coordinates (array components). For example, our universe has r = 4, a dimension array size of 2^8, and a $2^8 = 256$ dimension HyperSphere generated from 2^7 fractal curves.

5.3 Second Kind HyperCosmos Spaces' HyperSpheres

`An r dimension 2nd Kind HyperCosmos space and universe has a corresponding

$$2^{r+3} \text{ dimension HyperSphere} \tag{5.7}$$

and a dimension array containing 2^{r+3} dimensions. Its HyperSphere is generated by 2^{r+2} fractal curves.

5.4 A Child Universe Space-time is Also a HyperSphere

The set of dimensions of a child universe's space-time can also be viewed as generated by a HyperSphere composed of a set of r/2 fractal curves. Thus our 4D universe, if it were a child universe, would have a space-time HyperSphere of 4 dimensions with the HyperSphere generated by 2 fractal curves.

5.5 Conclusion

We conclude from the above discussion that a HyperSphere view of child universes is possible. It presents a new approach to understanding the ultimate origin of child universes. In previous books we have considered quantum field theories of inter-universe interactions and child universe generation. Together with HyperSphere Generation we have a new perspective on universe origin and dynamics!

6. Entropy and The Role of a Prime Mover

We have seen a thorough development of the structure of the Cosmos from a mathematical point in accordance with the entropic generation of the Cosmos spaces dimension arrays from a *single* Penultimate Unification space (non-zero) dimension. The rise to a high entropy value signals a direct, perhaps? unequivocal growth mechanism.

The rise in entropy signals the growth in the number of possible dimension array microstates. The rise in entropy in Cosmic processes parallels a Chemical transition from initial molecules to a set of reaction products.

As a result we now see a clear reason for the generation of Cosmos structure. The "Why?" question raised in our previous study is now answered by entropy. We have reasons for Cosmos structure based on the infinitesimal – a point or dimension and on inexorable entropy. The approach based on the growth of a Mathematical point also has an entropic justification. All follows from the infinitesimal!

These studies are aimed to provide illumination on the separation of the Physical from the Spiritual. Once "started" the evolution from the point to the Cosmos structure and the Parent universe is direct and unequivocally Physical.

The description that we have developed gives a clear picture of the interrelations and the consequences of the Extended Master Equation and entropy. The description is sequential. The evolution is instantaneous. The result is eternal. Events within universes are temporal.

We can see that the "start" of Cosmos structure requires something beyond the Physical. We view it as the action of a Prime Mover.[25] We can see that the process of Cosmos structure "creation" is determined by Physical considerations. We can see events within universes – being dynamic in nature – maybe Physical or the result of Prime Mover actions.

---

[25] Prime Mover is often used to designate God in Philosophy.

7. Why?

Round and round in a widening gyre,[26]
We come to see the Cosmic fire.

The question of *WHY* raised in a previous book opens the consideration of ultimate origins. We have seen that the Cosmos could have begun with a mathematical point. We have also seen that the Cosmos, equivalently, might have begun with a single non-zero dimension, which also might be viewed as a point.

The question that we face is the motivation, if there is one, for the process of Cosmos generation, for the process to begin. The question of motivation brings us to the further question: Whose motivation?

We can view the answer to this question as Physical or non-physical in nature. The great success in developing a Cosmos Theory, seated in Logic and Mathematics, with its enormous variety of features, with its direct contact with known experimental results such as the form of interaction symmetries, the strength of interactions, and the masses of fermions, LEADS to an overwhelming question: What agent (Prime Mover?) could produce such riches without the chaos of conflicts in Mathematics and Logic?

The answer, of course, is a matter of contention. We shall take a conservative approach based on what little we know.

We know that the agent is independent of the Cosmos. The Cosmos, once started, runs by itself through its inherent dynamics.

Being independent of the Cosmos – outside of the Cosmos so to speak – we must assume it is static in the sense that it does not change in itself and does not implement change in the Cosmos from the agent's vantage point *outside* the Cosmos. Its nature is thus *static* outside the Cosmos. It starts the Cosmos – although the start point is beyond the "past" of the (everlasting) Cosmos. The Cosmos being independent of time renders it ever present with no future end.

Given the above description one must ask: What is the agent's motivation? The only possible answer seems to be: the agent enables ("starts") the Cosmos so as to have universes within which it can act and transcend its static nature in a manner external to Cosmos structure.

The Cosmos, and its universes, embodies the dynamic nature of the Agent. Cosmos Theory brings the Cosmos to a point – and the point to the Agent.

---

[26] From The *Second Coming* by W. B. Yeats.

REFERENCES

Achiever, N. I., Frank, A. H. (tr), 1962, *The Calculus of Variations* (Blaisdell Publishing, New York, 1962).
Bjorken, J. D., Drell, S. D., 1964, *Relativistic Quantum Mechanics* (McGraw-Hill, New York, 1965).
Bjorken, J. D., Drell, S. D., 1965, *Relativistic Quantum Fields* (McGraw-Hill, New York, 1965).
Blaha, S., 1995, *C++ for Professional Programming* (International Thomson Publishing, Boston, 1995).
_______, 1998, *Cosmos and Consciousness* (Pingree-Hill Publishing, Auburn, NH, 1998 and 2002).
_______, 2002, *A Finite Unified Quantum Field Theory of the Elementary Particle Standard Model and Quantum Gravity Based on New Quantum Dimensions™ & a New Paradigm in the Calculus of Variations* (Pingree-Hill Publishing, Auburn, NH, 2002).
_______, 2004, *Quantum Big Bang Cosmology: Complex Space-time General Relativity, Quantum Coordinates™ Dodecahedral Universe, Inflation, and New Spin 0, ½, 1 & 2 Tachyons & Imagyons* (Pingree-Hill Publishing, Auburn, NH, 2004).
______, *2005a, Quantum Theory of the Third Kind: A New Type of Divergence-free Quantum Field Theory Supporting a Unified Standard Model of Elementary Particles and Quantum Gravity based on a New Method in the Calculus of Variations* (Pingree-Hill Publishing, Auburn, NH, 2005).
______, 2005b, *The Metatheory of Physics Theories, and the Theory of Everything as a Quantum Computer Language* (Pingree-Hill Publishing, Auburn, NH, 2005).
______, 2005c, *The Equivalence of Elementary Particle Theories and Computer Languages: Quantum Computers, Turing Machines, Standard Model, Superstring Theory, and a Proof that Gödel's Theorem Implies Nature Must Be Quantum* (Pingree-Hill Publishing, Auburn, NH, 2005).
______, 2006a, *The Foundation of the Forces of Nature* (Pingree-Hill Publishing, Auburn, NH, 2006).
______, 2006b, *A Derivation of ElectroWeak Theory based on an Extension of Special Relativity; Black Hole Tachyons; & Tachyons of Any Spin.* (Pingree-Hill Publishing, Auburn, NH, 2006).
_______, 2007a, *Physics Beyond the Light Barrier: The Source of Parity Violation, Tachyons, and A Derivation of Standard Model Features* (Pingree-Hill Publishing, Auburn, NH, 2007).
______, 2007b, *The Origin of the Standard Model: The Genesis of Four Quark and Lepton Species, Parity Violation, the ElectroWeak Sector, Color SU(3), Three Visible Generations of Fermions, and One Generation of Dark Matter with Dark Energy* (Pingree-Hill Publishing, Auburn, NH, 2007).
______, *2008a, A Direct Derivation of the Form of the Standard Model From GL(16) (Pingree-Hill Publishing, Auburn, NH, 2008).*
______, 2008b, *A Complete Derivation of the Form of the Standard Model With a New Method to Generate Particle Masses Second Edition* (Pingree-Hill Publishing, Auburn, NH, 2008)
______, 2009, *The Algebra of Thought & Reality: The Mathematical Basis for Plato's Theory of Ideas, and Reality Extended to Include A Priori Observers and Space-Time Second Edition* (Pingree-Hill Publishing, Auburn, NH, 2009).
______, 2009a, *Bright Stars, Bright Universe* (Pingree-Hill Publishing, Auburn, NH, 2008)
______, 2010a, *Operator Metaphysics: A New Metaphysics Based on a New Operator Logic and a New Quantum Operator Logic that Lead to a Mathematical Basis for Plato's Theory of Ideas and Reality* (Pingree-Hill Publishing, Auburn, NH, 2010).
______, 2010b, *The Standard Model's Form Derived from Operator Logic, Superluminal Transformations and GL(16)* (Pingree-Hill Publishing, Auburn, NH, 2010).
______, 2010c, *SuperCivilizations: Civilizations as Superorganisms* (McMann-Fisher Publishing, Auburn, NH, 2010).
______, 2011a, *21st Century Natural Philosophy Of Ultimate Physical Reality* (McMann-Fisher Publishing, Auburn, NH, 2011).
______, 2011b, *All the Universe! Faster Than Light Tachyon Quark Starships & Particle Accelerators with the LHC as a Prototype Starship Drive Scientific Edition* (Pingree-Hill Publishing, Auburn, NH, 2011).
______, 2011c, *From Asynchronous Logic to The Standard Model to Superflight to the Stars* (Blaha Research, Auburn, NH, 2011).

______, 2012a, *From Asynchronous Logic to The Standard Model to Superflight to the Stars volume 2: Superluminal CP and CPT, U(4) Complex General Relativity and The Standard Model, Complex Vierbein General Relativity, Kinetic Theory, Thermodynamics* (Blaha Research, Auburn, NH, 2012).
______, 2012b, *Standard Model Symmetries, And Four And Sixteen Dimension Complex Relativity; The Origin Of Higgs Mass Terms* (Blaha Reasearch, Auburn, NH, 2012).
______, 2013a, *Multi-Stage Space Guns, Micro-Pulse Nuclear Rockets, and Faster-Than-Light Quark-Gluon Ion Drive Starships* (Blaha Research, Auburn, NH, 2013).
______, 2013b, *The Bridge to Dark Matter; A New Sibling Universe; Dark Energy; Inflatons; Quantum Big Bang; Superluminal Physics; An Extended Standard Model Based on Geometry* (Blaha Reasearch, Auburn, NH, 2013).
______, 2014a, *Universes and Megaverses: From a New Standard Model to a Physical Megaverse; The Big Bang; Our Sibling Universe's Wormhole; Origin of the Cosmological Constant, Spatial Asymmetry of the Universe, and its Web of Galaxies; A Baryonic Field between Universes and Particles; Megaverse Extended Wheeler-DeWitt Equation* (Blaha Reasearch, Auburn, NH, 2014).
______, 2014b, *All the Megaverse! Starships Exploring the Endless Universes of the Cosmos Using the Baryonic Force* (Blaha Research, Auburn, NH, 2014).
______, 2014c, *All the Megaverse! II Between Megaverse Universes: Quantum Entanglement Explained by the Megaverse Coherent Baryonic Radiation Devices – PHASERs Neutron Star Megaverse Slingshot Dynamics Spiritual and UFO Events, and the Megaverse Microscopic Entry into the Megaverse* (Blaha Research, Auburn, NH, 2014).
______, 2015a, *PHYSICS IS LOGIC PAINTED ON THE VOID: Origin of Bare Masses and The Standard Model in Logic, U(4) Origin of the Generations, Normal and Dark Baryonic Forces, Dark Matter, Dark Energy, The Big Bang, Complex General Relativity, A Megaverse of Universe Particles* (Blaha Research, Auburn, NH, 2015).
______, 2015b, *PHYSICS IS LOGIC Part II: The Theory of Everything, The Megaverse Theory of Everything, U(4)⊗U(4) Grand Unified Theory (GUT), Inertial Mass = Gravitational Mass, Unified Extended Standard Model and a New Complex General Relativity with Higgs Particles, Generation Group Higgs Particles* (Blaha Research, Auburn, NH, 2015).
____, 2015c, *The Origin of Higgs ("God") Particles and the Higgs Mechanism: Physics is Logic III, Beyond Higgs – A Revamped Theory With a Local Arrow of Time, The Theory of Everything Enhanced, Why Inertial Frames are Special, Universes of the Mind* (Blaha Research, Auburn, NH, 2015).
______, 2015d, *The Origin of the Eight Coupling Constants of The Theory of Everything: U(8) Grand Unified Theory of Everything (GUTE), S^8 Coupling Constant Symmetry, Space-Time Dependent Coupling Constants, Big Bang Vacuum Coupling Constants, Physics is Logic IV* (Blaha Research, Auburn, NH, 2015).
______, 2016a, *New Types of Dark Matter, Big Bang Equipartition, and A New U(4) Symmetry in the Theory of Everything: Equipartition Principle for Fermions, Matter is 83.33% Dark, Penetrating the Veil of the Big Bang, Explicit QFT Quark Confinement and Charmonium, Physics is Logic V* (Blaha Research, Auburn, NH, 2016).
______, 2016b, *The Periodic Table of the 192 Quarks and Leptons in The Theory of Everything: The U(4) Layer Group, Physics is Logic VI* (Blaha Research, Auburn, NH, 2016).
______, 2016c, *New Boson Quantum Field Theory, Dark Matter Dynamics, Dark Matter Fermion Layer Mixing, Genesis of Higgs Particles, New Layer Higgs Masses, Higgs Coupling Constants, Non-Abelian Higgs Gauge Fields, Physics is Logic VII* (Blaha Research, Auburn, NH, 2016).
______, 2016d, *Unification of the Strong Interactions and Gravitation: Quark Confinement Linked to Modified Short-Distance Gravity; Physics is Logic VIII* (Blaha Research, Auburn, NH, 2016).
______, 2016e, *MoND: Unification of the Strong Interactions and Gravitation II, Quark Confinement Linked to Large-Scale Gravity, Physics is Logic IX* (Blaha Research, Auburn, NH, 2016).
______, 2016f, *CQ Mechanics: A Unification of Quantum & Classical Mechanics, Quantum/Semi-Classical Entanglement, Quantum/Classical Path Integrals, Quantum/Classical Chaos* (Blaha Research, Auburn, NH, 2016).
______, 2016g, *GEMS Unified Gravity, ElectroMagnetic and Strong Interactions: Manifest Quark Confinement, A Solution for the Proton Spin Puzzle, Modified Gravity on the Galactic Scale* (Pingree Hill Publishing, Auburn, NH, 2016).
______, 2016h, *Unification of the Seven Boson Interactions based on the Riemann-Christoffel Curvature Tensor* (Pingree Hill Publishing, Auburn, NH, 2016).
______, 2017a, *Unification of the Eleven Boson Interactions based on 'Rotations of Interactions'* (Pingree Hill Publishing, Auburn, NH, 2017).
______, 2017b, *The Origin of Fermions and Bosons, and Their Unification* (Pingree Hill Publishing, Auburn, NH, 2017).
______, 2017c, *Megaverse: The Universe of Universes* (Pingree Hill Publishing, Auburn, NH, 2017).
______, 2017d, *SuperSymmetry and the Unified SuperStandard Model* (Pingree Hill Publishing, Auburn, NH, 2017).

______, 2017e, *From Qubits to the Unified SuperStandard Model with Embedded SuperStrings: A Derivation* (Pingree Hill Publishing, Auburn, NH, 2017).
______, 2017f, *The Unified SuperStandard Model in Our Universe and the Megaverse: Quarks, ... ,* (Pingree Hill Publishing, Auburn, NH, 2017).
______, 2018a, *The Unified SuperStandard Model and the Megaverse SECOND EDITION A Deeper Theory based on a New Particle Functional Space that Explicates Quantum Entanglement Spookiness (Volume 1)* (Pingree Hill Publishing, Auburn, NH, 2018).
______, 2018b, *Cosmos Creation: The Unified SuperStandard Model, Volume 2, SECOND EDITION* (Pingree Hill Publishing, Auburn, NH, 2018).
______, 2018c, *God Theory (*Pingree Hill Publishing, Auburn, NH, 2018).
______, 2018d, *Immortal Eye: God Theory: Second Edition* (Pingree Hill Publishing, Auburn, NH, 2018).
______, 2018e, *Unification of God Theory and Unified SuperStandard Model THIRD EDITION* (Pingree Hill Publishing, Auburn, NH, 2018).
______, 2019a, *Calculation of: QED α = 1/137, and Other Coupling Constants of the Unified SuperStandard Theory* (Pingree Hill Publishing, Auburn, NH, 2019).
______, 2019b, *Coupling Constants of the Unified SuperStandard Theory SECOND EDITION* (Pingree Hill Publishing, Auburn, NH, 2019).
______, 2019c, *New Hybrid Quantum Big_Bang–Megaverse_Driven Universe with a Finite Big Bang and an Increasing Hubble Constant* (Pingree Hill Publishing, Auburn, NH, 2019).
______, 2019d, *The Universe, The Electron and The Vacuum* (Pingree Hill Publishing, Auburn, NH, 2019).
______, 2019e, *Quantum Big Bang – Quantum Vacuum Universes (Particles)* (Pingree Hill Publishing, Auburn, NH, 2019).
______, 2019f, *The Exact QED Calculation of the Fine Structure Constant Implies ALL 4D Universes have the Same Physics/Life Prospects* (Pingree Hill Publishing, Auburn, NH, 2019).
______, 2019g, *Unified SuperStandard Theory and the SuperUniverse Model: The Foundation of Science* (Pingree Hill Publishing, Auburn, NH, 2019).
______, 2020a, *Quaternion Unified SuperStandard Theory (The QUeST) and Megaverse Octonion SuperStandard Theory (MOST)* (Pingree Hill Publishing, Auburn, NH, 2020).
______, 2020b, *United Universes Quaternion Universe - Octonion Megaverse* (Pingree Hill Publishing, Auburn, NH, 2020).
______, 2020c, *Unified SuperStandard Theories for Quaternion Universes & The Octonion Megaverse* (Pingree Hill Publishing, Auburn, NH, 2020).
______, 2020d, *The Essence of Eternity: Quaternion & Octonion SuperStandard Theories* (Pingree Hill Publishing, Auburn, NH, 2020).
______, 2020e, *The Essence of Eternity II* (Pingree Hill Publishing, Auburn, NH, 2020).
______, 2020f, *A Very Conscious Universe* (Pingree Hill Publishing, Auburn, NH, 2020).
______, 2020g, *Hypercomplex Universe* (Pingree Hill Publishing, Auburn, NH, 2020).
______, 2020h, *Beneath the Quaternion Universe* (Pingree Hill Publishing, Auburn, NH, 2020).
______, 2020i, *Why is the Universe Real? From Quaternion & Octonion to Real Coordinates* (Pingree Hill Publishing, Auburn, NH, 2020).
______, 2020j, *The Origin of Universes: of Quaternion Unified SuperStandard Theory (QUeST); and of the Octonion Megaverse (UTMOST)* (Pingree Hill Publishing, Auburn, NH, 2020).
______, 2020k, *The Seven Spaces of Creation: Octonion Cosmology* (Pingree Hill Publishing, Auburn, NH, 2020).
______, 2020l, *From Octonion Cosmology to the Unified SuperStandard Theory of Particles* (Pingree Hill Publishing, Auburn, NH, 2020).
______, 2021a, *Pioneering the Cosmos* (Pingree Hill Publishing, Auburn, NH, 2021).
______, 2021b, *Pioneering the Cosmos II* (Pingree Hill Publishing, Auburn, NH, 2021).
______, 2021c, *Beyond Octonion Cosmology* (Pingree Hill Publishing, Auburn, NH, 2021).
______, 2021d, *Universes are Particles* (Pingree Hill Publishing, Auburn, NH, 2021).
______, 2021e, *Octonion-like dna-based life, Universe expansion is decay, Emerging New Physics* (Pingree Hill Publishing, Auburn, NH, 2021).
______, 2021f, *The Science of Creation New Quantum Field Theory of Spaces* (Pingree Hill Publishing, Auburn, NH, 2021).
______, 2021g, *Quantum Space Theory With Application to Octonion Cosmology & Possibly To Fermionic Condensed Matter* (Pingree Hill Publishing, Auburn, NH, 2021).
______, 2021h, *21st Century Natural Philosophy of Octonion Cosmology , and Predestination, Fate, and Free Will* (Pingree Hill Publishing, Auburn, NH, 2021).

______, 2021i, *Beyond Octonion Cosmology II : Origin of the Quantum; A New Generalized Field Theory (GiFT); A Proof of the Spectrum of Universes; Atoms in Higher Universes* (Pingree Hill Publishing, Auburn, NH, 2021).
______, 2021j, *Integration of General Relativity and Quantum Theory: Octonion Cosmology, GiFT, Creation/Annihilation Spaces CASe, Reduction of Spaces to a Few Fermions and Symmetries in Fundamental Frames* (Pingree Hill Publishing, Auburn, NH, 2021).
______, 2022a, *New View of Octonion Cosmology Based on the Unification of General Relativity and Quantum Theory* (Pingree Hill Publishing, Auburn, NH, 2022).
______, 2022b, *The Dust Beneath Hypercomplex Cosmology* (Pingree Hill Publishing, Auburn, NH, 2022).
______, 2022c, *Passing Through Nature to Eternity: ProtoCosmos, HyperCosmos, Unified SuperStandard Theory* (Pingree Hill Publishing, Auburn, NH, 2022).
______, 2022d, *HyperCosmos Fractionation and Fundamental Reference Frame Based Unification: Particle Inner Space Basis of Parton and Dual Resonance Models* (Pingree Hill Publishing, Auburn, NH, 2022).
______, 2022e, *A New UniDimension ProtoCosmos and SuperString F-Theory Relation to the HyperCosmos* (Pingree Hill Publishing, Auburn, NH, 2022).
______, 2022f, *The Cosmic Panorama: ProtoCosmos, HyperCosmos, Unified SuperStandard Theory (UST) Derivation* (Pingree Hill Publishing, Auburn, NH, 2022).
______, 2022g, *Ultimate Origin: ProtoCosmos and HyperCosmos* (Pingree Hill Publishing, Auburn, NH, 2022).
______, 2023a, *UltraUnification and the Generation of the Cosmos* (Pingree Hill Publishing, Auburn, NH, 2023).
______, 2023b, *God and and Cosmos Theory* (Pingree Hill Publishing, Auburn, NH, 2023).
______, 2023c, *A New Completely Geometric SU(8) Cosmos Theory; New PseudoFermion Fields; Fibonacci-like Dimension Arrays; Ramsey Number Approximation* (Pingree Hill Publishing, Auburn, NH, 2023).
______, 2023d, *Newton's Apple is Now the Fermion* (Pingree Hill Publishing, Auburn, NH, 2023).
______, 2023e,*Cosmos Theory: The Sub-Particle Gambol Model* (Pingree Hill Publishing, Auburn, NH, 2023).
______, 2024a, *Cosmos-Universe-Particle-Gambol Theory* (Pingree Hill Publishing, Auburn, NH, 2024).
______, 2024b, *Fractal Cosmos Theory* (Pingree Hill Publishing, Auburn, NH, 2024).
______, 2024c, *Fractal Cosmic Curve: Tensor-Based CosmosTheory* (Pingree Hill Publishing, Auburn, NH, 2024).
______, 2024d, *The Eternal Form of Cosmos Theory* (Pingree Hill Publishing, Auburn, NH, 2024).
______, 2024e, *The Eternal Form of Cosmos Theory Third Edition* (Pingree Hill Publishing, Auburn, NH, 2024).
______, 2024f, *Fundamental Constants of Cosmos Theory and The Standard Model* (Pingree Hill Publishing, Auburn, NH, 2024).
______, 2024g, *Quark, Lepton, W and Z Masses of Cosmos Theory and The Standard Model* (Pingree Hill Publishing, Auburn, NH, 2024).
______, 2024h, *Geometric Cosmos Geometric Universe* (Pingree Hill Publishing, Auburn, NH, 2024).
______, 2024i, *Particles and Universes of Cosmos Theory* (Pingree Hill Publishing, Auburn, NH, 2024).
______, 2024j, *Unification of the Subluminal and the Superluminal in Cosmos Theory* (Pingree Hill Publishing, Auburn, NH, 2024).
______, 2024k, *The Dawn of Dynamic Cosmos Dimension Arrays* (Pingree Hill Publishing, Auburn, NH, 2024).
______, 2024ℓ, *Structure and Dynamics of Cosmos Theory and the Unified SuperStandard Theory* (Pingree Hill Publishing, Auburn, NH, 2024).
______, 2025a, *Black Holes, White Holes, and Superluminal Starship* (Pingree Hill Publishing, Auburn, NH, 2025).
______, 2025b, *The Sequence Theory Foundation of Cosmos Theory and Gambol Theory* (Pingree Hill Publishing, Auburn, NH, 2025).
____, 2025c, *BEYOND TIME AND SPACE Fermion Strata & Consistency Conditions for Sequences, Cosmos Spaces and Gambols* (Pingree Hill Publishing, Auburn, NH, 2025).
____, 2025d, *The C o s m o s as a Physics Organism* (Blaha Research, Auburn, NH, 2025).
____, 2025e, *The Root of Reality* (Blaha Research, Auburn, NH, 2025).
____, 2025f, *The Fractal Cosmos* (Blaha Research, Auburn, NH, 2025).
____, 2025g, *The Fractal Cosmos II* (Blaha Research, Auburn, NH, 2025).
____, 2025h, *The Fractal Cosmos III* (Blaha Research, Auburn, NH, 2025).
____, 2025i, *Quantum Coalescent Cosmos Fractal Cosmos IV* (Blaha Research, Auburn, NH, 2025).
Eddington, A. S., 1952, *The Mathematical Theory of Relativity* (Cambridge University Press, Cambridge, U.K., 1952).
Fant, Karl M., 2005, *Logically Determined Design: Clockless System Design With NULL Convention Logic* (John Wiley and Sons, Hoboken, NJ, 2005).
Feinberg, G. and Shapiro, R., 1980, *Life Beyond Earth: The Intelligent Earthlings Guide to Life in the Universe* (William Morrow and Company, New York, 1980).

Gelfand, I. M., Fomin, S. V., Silverman, R. A. (tr), 2000, *Calculus of Variations* (Dover Publications, Mineola, NY, 2000).
Giaquinta, M., Modica, G., Souchek, J., 1998, *Cartesian Coordinates in the Calculus of Variations* Volumes I and II (Springer-Verlag, New York, 1998).
Giaquinta, M., Hildebrandt, S., 1996, *Calculus of Variations* Volumes I and II (Springer-Verlag, New York, 1996).
Gradshteyn, I. S. and Ryzhik, I. M., 1965, *Table of Integrals, Series, and Products* (Academic Press, New York, 1965).
Heitler, W., 1954, *The Quantum Theory of Radiation* (Claendon Press, Oxford, UK, 1954).
Huang, Kerson, 1992, *Quarks, Leptons & Gauge Fields 2nd Edition* (World Scientific Publishing Company, Singapore, 1992).
Jost, J., Li-Jost, X., 1998, *Calculus of Variations* (Cambridge University Press, New York, 1998).
Kaku, Michio, 1993, *Quantum Field Theory*, (Oxford University Press, New York, 1993).
Kirk, G. S. and Raven, J. E., 1962, *The Presocratic Philosophers* (Cambridge University Press, New York, 1962).
Landau, L. D. and Lifshitz, E. M., 1987, *Fluid Mechanics 2nd Edition*, (Pergamon Press, Elmsford, NY, 1987).
Rescher, N., 1967, *The Philosophy of Leibniz* (Prentice-Hall, Englewood Cliffs, NJ, 1967).
Riesz, Frigyes and Sz.-Nagy, Béla, 1990, *Functional Analysis* (Dover Publications, New York, 1990).
Sakurai, J. J., 1964, *Invariance Principles and Elementary Particles* (Princeton University Press, Princeton, NJ, 1964).
Weinberg, S., 1972, *Gravitation and Cosmology* (John Wiley and Sons, New York, 1972).
Weinberg, S., 1995, *The Quantum Theory of Fields Volume I* (Cambridge University Press, New York, 1995).

INDEX

About the Author

Stephen Blaha is a well-known Physicist and Man of Letters with interests in Science, Society and civilization, the Arts, and Technology. He had an Alfred P. Sloan Foundation scholarship in college. He received his Ph.D. in Physics from Rockefeller University. He has served on the faculties of several major universities. He was also a Member of the Technical Staff at Bell Laboratories, a manager at the Boston Globe Newspaper, a Director at Wang Laboratories, and President of Blaha Software Inc. and of Janus Associates Inc. (NH).

Among other achievements he was a co-discoverer of the "r potential" for heavy quark binding developing the first (and still the only demonstrable) non-Aeolian gauge theory with an "r" potential; first suggested the existence of topological structures in superfluid He-3; first proposed Yang-Mills theories would appear in condensed matter phenomena with non-scalar order parameters; first developed a grammar-based formalism for quantum computers and applied it to elementary particle theories; first developed a new form of quantum field theory without divergences (thus solving a major 60 year old problem that enabled a unified theory of the Standard Model and Quantum Gravity without divergences to be developed); first developed a formulation of complex General Relativity based on analytic continuation from real space-time; first developed a generalized non-homogeneous Robertson-Walker metric that enabled a quantum theory of the Big Bang to be developed without singularities at t = 0; first generalized Cauchy's theorem and Gauss' theorem to complex, curved multi-dimensional spaces; received Honorable Mention in the Gravity Research Foundation Essay Competition in 1978; first developed a physically acceptable theory of faster-than-light particles; first derived a composition of extremums method in the Calculus of Variations; first quantitatively suggested that inflationary periods in the history of the universe were not needed; first proved Gödel's Theorem implies Nature must be quantum; provided a new alternative to the Higgs Mechanism, and Higgs particles, to generate masses; first showed how to resolve logical paradoxes including Gödel's Undecidability Theorem by developing Operator Logic and Quantum Operator Logic; first developed a quantitative harmonic oscillator-like model of the life cycle, and interactions, of civilizations; first showed how equations describing superorganisms also apply to civilizations. A recent book shows his theory applies successfully to the past 14 years of history and to *new* archaeological data on Andean and Mayan civilizations as well as Early Anatolian and Egyptian civilizations.

He first developed an axiomatic derivation of the form of The Standard Model from geometry – space-time properties – The Unified SuperStandard Model. It unifies all the known forces of Nature. It also has a Dark Matter sector that includes a Dark ElectroWeak sector with Dark doublets and Dark gauge interactions. It uses quantum coordinates to remove infinities that crop up in most interacting quantum field theories and additionally to remove the infinities that appear in the Big Bang and generate inflationary growth of the universe. It shows gravity has a MOND-like form without sacrificing Newton's Laws. It relates the interactions of the MOND-like sector of gravity with the r-potential of Quark Confinement. The axioms of the theory lead to the question of their origin. We suggest in the preceding edition of this book it can be attributed to an entity with God-like properties. We explore these properties in "God Theory" and show they predict that the Cosmos exists forever although individual universes (or incarnations of our universe) "come and go." Several other important results emerge from God Theory such a functionally triune God. The Unified SuperStandard Theory has many other important parts described in the Current Edition of *The Unified SuperStandard Theory* and expanded in subsequent volumes.

Blaha has had a major impact on a succession of elementary particle theories: his Ph.D. thesis (1970), and papers, showed that quantum field theory calculations to all orders in ladder approximations could not give scaling deep inelastic electron-nucleon scattering. He later showed the eigenvalue equation for the fine structure constant α in Johnson-Baker-Willey QED had a zero at $\alpha = 1$ not 1/137 by solving the Schwinger-Dyson equations to all orders in an approximation that agreed with exact results to 4th order in α thus ending interest in this theory. In 1979 at Professor Ken Johnson's (MIT) suggestion he calculated the proton-neutron mass difference in the MIT bag model and found the result had the wrong sign reducing interest in the bag model. These results all appear in Physical Review papers. In the 2000's he repeatedly pointed out the shortcomings of SuperString theory and showed that The Standard Model's form could be derived from space-time geometry by an extension of Lorentz transformations to faster than light transformations. This deeper space-time basis greatly increases the possibility that it is part of THE fundamental theory. Recently, Blaha showed that the Weak interactions differed significantly from the Strong, electromagnetic and gravitation interactions in important respects while these interactions had similar features, and suggested that ElectroWeak theory, which is essentially a glued union of the Weak interactions and Electromagnetism, possibly modulo unknown Higgs particle features, be replaced by a unified theory of the other interactions combined with a stand-alone Weak interaction theory. Blaha also showed that, if Charmonium calculations are taken seriously, the Strong interaction coupling constant is only a factor of five larger than the electromagnetic coupling constant, and thus Strong interaction perturbation theory would make sense and yield physically meaningful results.

In graduate school (1965-71) he wrote substantial papers in elementary particles and group theory: The Inelastic E- P Structure Functions in a Gluon Model. Phys. Lett. B40:501-502,1972; Deep-Inelastic E-P Structure Functions In A Ladder Model With Spin 1/2 Nucleons, Phys.Rev. D3:510-523,1971; Continuum Contributions To The Pion Radius, Phys. Rev. 178:2167-2169,1969; Character Analysis of U(N) and SU(N), J. Math. Phys. 10, 2156 (1969); and The Calculation of the Irreducible Characters of the Symmetric Group in Terms of the Compound Characters, (Published as Blaha's Lemma in D. E. Knuth's book: *The Art of Computer Programming Vols. 1 – 4*).

In the early 1980's Blaha was also a pioneer in the development of UNIX for financial, scientific and Internet applications: benchmarked UNIX versions showing that block size was critical for UNIX performance, developing financial modeling software, starting database benchmarking comparison studies, developing Internet-like UNIX networking (1982) and developing a hybrid shell programming technique (1982) that was a precursor to the PERL programming language. He was also the manager of the AT&T ten-year future products development database. His work helped lead to commercial UNIX on computers such as Sun Micros, IBM AIX minis, and Apple computers.

In the 1980's he pioneered the development of PC Desktop Publishing on laser printers and was nominated for three "Awards for Technical Excellence" in 1987 by PC Magazine for PC software products that he designed and developed.

Recently he has developed a theory of Megaverses – actual universes of which our universe is one – with quantum particle-like properties based on the Wheeler-DeWitt equation of Quantum Gravity. He has developed a theory of a baryonic force, which had been conjectured many years ago, and estimated the strength of the force based on discrepancies in measurements of the gravitational constant G. This force, operative in D-dimensional space, can be used to escape from our universe in "uniships" which are the equivalent of the faster-than-light starships proposed in the author's earlier books. Thus travel to other universes, as well as to other stars is possible.

Blaha also considered the complexified Wheeler-DeWitt equation and showed that its limitation to real-valued coordinates and metrics generated a Cosmological Constant in the Einstein equations.

The author has also recently written a series of books on the serious problems of the United States and their solution as well as a book on the decline of Mankind that will follow from current social and genetic trends in Mankind.

In the past twenty years Dr. Blaha has written over 80 books on a wide range of topics. Some recent major works are: *From Asynchronous Logic to The Standard Model to Superflight to the Stars, All the Universe!, SuperCivilizations: Civilizations as Superorganisms, America's Future: an Islamic Surge, ISIS, al Qaeda, World Epidemics, Ukraine, Russia-China Pact, US Leadership Crisis, The Rises and Falls of Man – Destiny – 3000 AD: New Support for a Superorganism MACRO-THEORY of CIVILIZATIONS From CURRENT WORLD TRENDS and NEW Peruvian, Pre-Mayan, Mayan, Anatolian, and Early Egyptian Data, with a Projection to 3000 AD,* and *Mankind in Decline: Genetic Disasters, Human-Animal Hybrids, Overpopulation, Pollution, Global Warming, Food and Water Shortages, Desertification, Poverty, Rising Violence, Genocide, Epidemics, Wars, Leadership Failure.*

He has taught approximately 4,000 students in undergraduate, graduate, and postgraduate corporate education courses primarily in major universities, and large companies and government agencies.

He developed a quantum theory, The Unified SuperStandard Theory (UST), which describes elementary particles in detail without the difficulties of conventional quantum field theory. He found that the internal symmetries of this theory could be exactly derived from an octonion theory called QUeST. He further found that another octonion theory (UTMOST) describes the Megaverse. It can hold QUeST universes such as our own universe. It has an internal symmetry structure which is a superset of the QUeST internal symmetries.

Recently he developed Octonion Cosmology. He replaced it with HypcrCosmos theory, which has significantly better features. He developed a fractionalization process for dimensions, particles and symmetry groups. He also described transformation that reduced particles and dimensions to a far more compact form. He also developed a precursor theory ProtoCosmos that leads to the HyperCosmos.

The author showed that space-time and Internal Symmetries can be unified in any of the ten HyperCosmos spaces in their associated HyperUnification spaces. The combined set of HyperUnification spaces enable all HyperCosmos dimensions to be obtained by a General Relativistic transformation from one primordial dimension in the 42 space-time dimension unified HyperUnification space.

At present the author devel;oped the Cosmos Theory that incorporates ProtoCosmos Theory, HyperCosmos Theory, Limos Theory, Second Kind HyperCosmos Theory and HyperUnification Spaces. He has introduced PseudoFermion wave functions and theory, He has related Cosmos Theory to Regge trajectories of spaces, parton theory, Veneziano amplitudes, Fibonacci numbers and Ramsey numbers. He has calculated an approximation to the difficult R(n,n) Ramsey numbers.

He has developed a Gambol Model that successfully accounts for e-p deep inelastic scattering, fundamental particle resonances, hadron scattering, and the inner structure of particles based on confinement through Casimir forces of ideal gambol gases. The Gambol Planckian Distribution was derived.

He has applied the Gambol Model to particles, universes, and the Cosmos of universes. He showed that the Cosmos may have a distribution of 23 universes corresponding to various Cosmos spaces.

Recently he showed that Cosmos Theory follows from the number of independent asymmetric tensors in a dimension r. He also showed the close parallel between the form of γ-matrices and Cosmos Theory dimension arrays. The closeness suggested that dimension arrays have the same importance as γ-matrices for fermions.

He demonstrated that the pressure of fermions within a space of dimension r balances the Casimir vacuum energy force for 18 dimensions. He showed that $2e\pi = 17.02$ marks the critical point where pressure balances Casimir force, which implies $r = 18$ is the highest dimension Physical Cosmos space. The dimension $2e\pi$ appears to set the approximate dimension for Cosmos spaces with dimension array size $2^{r+4} \cong (17.02/8)^{r+4}. \cong (e\pi/4)^{r+4}. \cong 2.13^{r+4}$.

Now he has found the sequences of Coupling Constant values and fermion masses in the Standard Model and UST. He unified the Cosmos spaces spectrum, the coupling constant spectrum and the fermion mass sequences in a fundamental scaling differential equation.

The fractal basis of Cosmos Theory is now well explored.